DESIGN OF MEDICAL ELECTRONIC DEVICES

DESIGN OF MEDICAL ELECTRONIC DEVICES

Reinaldo Perez, PhD, PE
President
M. R. Research Inc.

ACADEMIC PRESS
An Elsevier Science Imprint

San Diego   San Francisco   New York
Boston   London   Sydney   Tokyo

This book is printed on acid-free paper. ∞

Copyright © 2002 by Academic Press

All rights reserved.
No part of this publication may be reproduced or transmitted in any form or by any means, electronic or mechanical, including photocopying, recording, or any information storage and retrieval system, without permission in writing from the publisher. Requests for permission to make copies of any part of the work should be mailed to the following address: Permissions Department, Harcourt, Inc., 6277 Sea Harbor Drive, Orlando, Florida 32887-6777.

ACADEMIC PRESS
An Elsevier Science Imprint
525 B Street, Suite 1900, San Diego, CA 92101-4495, USA
http://www.academicpress.com

Academic Press
Harcourt Place, 32 Jamestown Road, London, NW1 7BY, UK
http://www.academicpress.com

Library of Congress Catalog Number: 2001095334
ISBN: 0-12-550711-9

Printed in the United States of America

01 02 03 04 05 06 HP 9 8 7 6 5 4 3 2 1

CONTENTS

ACKNOWLEDGMENTS xi

Introduction 1

1 Proper Design of Power Subsystems in Medical Electronics 3

1.1 Electromagnetic Interference Requirements 3
1.2 Transient Voltage Protection 3
1.3 Electromagnetic Interference 8
    1.3.1 Electromagnetic Interference Requirement 8
1.4 Inrush Current Control 12
1.5 Soft Start 13
1.6 Overvoltage Protection 14
1.7 Undervoltage Protection 14
1.8 Overload Protection 16
1.9 Snubber Circuits 17
1.10 Output Filtering 19
1.11 Power Failure Warning 21
1.12 Flightback Switch Mode Power Supplies 22
1.13 Half-Bridge Flyback Converter 24

1.14 Forward Converter 26
1.15 High-Voltage Defibrillators 28
References 30

2 Fundamentals of Magnetic Resonance Imaging 33

2.1 Early History of Nuclear Magnetic Resonance 33
2.2 General Review of MRI 37
2.3 A More Detailed Overview of MRI 40
    2.3.1 *The Physics of Spin* 41
2.4 Magnetic Resonance Imaging Hardware Design 50
2.5 Pulsing and NMR Imaging 53
References 56

3 Particle Accelerator Design 57

3.1 Introduction 57
3.2 Heavy Particles 60
3.3 Particle Accelerators 60
3.4 Linear RF Accelerators 63
3.5 Particles Accelerated by a Magnetic Field 64
3.6 Synchrotrons 66
3.7 Accelerator Hardware 67
3.8 Magnetrons 69
3.9 Accelerator Architecture 70
    3.9.1 *Traveling Wave Accelerator System* 71
    3.9.2 *Standing Wave Accelerator* 71
References 74

4 Sensor Characteristics 75

4.1 Sensor Parameters 75
4.2 Physical Principles of Sensing 78
4.3 Sensor Interfacing 79
4.4 Driving Bridges 80
4.5 Signal-Conditioning Amplifiers 84
    4.5.1 *Noise* 84
4.6 Instrumentation Amplifiers 86
4.7 Chopper-Stabilized Amplifiers 88
4.8 Isolation Amplifier 89
4.9 Strain, Force, Pressure, and Flow Sensors 90
4.10 High-Impedance Sensors 93
4.11 High-Impedance Charge Output 95
4.12 Charge-Coupled Device Sensors 95
4.13 Position and Motion Sensors 98

        4.13.1   *Linear Variable Differential Transformers*   98
        4.13.2   *Hall Effect Magnetic Sensors*   100
        4.13.3   *Optical Encoders*   102
        4.13.4   *Accelerometers*   102
4.14 Temperature Sensors   104
        References   107

5 Data Acquisition   109

5.1   Introduction   109
5.2   Sample and Hold Conversion   110
5.3   Multichannel Acquisition   111
5.4   High-Speed Sampling in ADCs   112
5.5   Selection of Drive Amplifier for ADC Performance   116
5.6   Driving ADCs with Switched Capacitor Inputs   118
5.7   Gain Setting and Level Shifting   120
5.8   High-Speed Sampling ADC External Reference Voltage Generation   122
5.9   ADC Input Protection   123
5.10 Noise Considerations in High-Speed Sampling ADCs   124
5.11 Multichannel Applications for Data Acquisition Systems   128
5.12 External Protection of Amplifiers   131
5.13 High-Speed ADC Architectures   135
        5.13.1   *Basic Flash Converter Operation*   135
        5.13.2   *Driving Flash Converters*   135
        5.13.3   *Successive Approximation ADCs*   138
        5.13.4   *Subranging ADCs*   139
        References   141

6 Noise and Interference Issues in Analog Circuits   143

6.1   Basic Noise Calculation in Op-Amps   143
        6.1.1   *Thermal Noise*   144
6.2   Fundamental Op-Amp Specifications   146
6.3   Input Offset Voltage   151
6.4   The Noise Gain of Op-Amps   152
6.5   Slew Rate and Power Bandwidth of Op-Amps   152
6.6   Gain–Bandwidth Products   153
6.7   Internal Noise in Op-Amps   154
6.8   Noise Issues in High-Speed ADC Applications   157
6.9   Proper Power Supply Decoupling   159
6.10 Bypass Capacitors and Resonances   162
6.11 Usage of Two or More Bypass Capacitors   166
6.12 Designing Power Bus Rails in Power–Ground Planes for Noise Control   168
6.13 The Effect of Trace Resistance   171

6.14 ASIC Signal Integrity Issues (Ground Bounce)   176
6.15 Crosstalk through PC Card Pins   179
6.16 Parasitic Extraction and Verification Tools for ASIC   181
References   183

7 Hardware Approach to Digital Signal Processing   185

7.1 Discrete Fast Fourier Transform   185
7.2 Determining the Proper FFT Record Length   186
7.3 Coherent and Noncoherent Sampling   188
7.4 Coherent vs Noncoherent Sampling   189
7.5 Ultrasound Application   191
7.6 Discrete Time Sampling of Analog Signals   194
7.7 Digital Signal Processing Techniques   196
7.8 Finite Impulse Response Digital Filters   197
7.9 Infinite Impulse Response Digital Filters   200
7.10 Fast Fourier Transform   202
7.11 FFT Hardware Implementation   204
  7.11.1 DSP Hardware   205
  7.11.2 Arithmetic Logic Unit   207
  7.11.3 Multiplier–Accumulator   208
  7.11.4 Shifter   210
  7.11.5 Data Address Generators   211
  7.11.6 Program Sequencer   212
  7.11.7 Serial Ports   213
  7.11.8 System Interface   214
  7.11.9 Interfacing ADCs and DACs to Digital Signal Processors   214
  7.11.10 Parallel ADC-to-DSP Interface   217
  7.11.11 Parallel Interfacing to DSP Processors: Writing Data to Memory-Mapped DACs   220
  7.11.12 Parallel DAC-to-DSP Interface   220
  7.11.13 Serial Interfacing to DSP Processors   223
  7.11.14 Serial ADC-to-DSP Interface   225
  7.11.15 Serial DAC-to-DSP Interface   227
  7.11.16 Interfacing I–O Ports and CODECs to DSPs   229
  7.11.17 Serial versus Parallel DSP Interface Summary   233
7.12 Practical Use of DSP: DSP Helps the Hearing Impaired   234
References   236

8 Optical Sensors   237

8.1 Charge-Coupled Devices   237
  8.1.1 CCD Arrays   240
  8.1.2 Interline Transfer   240

8.2 Optical Fiber 242
    8.2.1 Classification and Features of Optical Fibers 244
8.3 Analysis of Optical Fibers 251
    8.3.1 The Step-Index Fiber 251
8.4 The Graded-Index Fiber 257
8.5 CT Scanners in Medicine 259
    8.5.1 Sectional Imaging 261
    8.5.2 Digital Imaging 263
8.6 The Endoscope 263
8.7 Digital X Rays 264
8.8 Medical Sensors from Fiber Optics 268
    8.8.1 Fiber Optics for Circulatory and Respiratory Systems 269
    References 273

INDEX 275

ACKNOWLEDGMENTS

I would like to use this opportunity to thank Madeline Kelly Reilly-Perez for her tireless efforts in typing this manuscript and producing many of the figures. Without her help this book would not have been possible.

■ INTRODUCTION

Most of the advances in medicine today are related to medical technology. On one hand, we have the pharmaceutical companies with their efforts to develop new drugs and treatments for many kinds of diseases. These efforts have been for the most part very lucrative since pharmaceutical and biotechnology companies have for many years enjoyed considerable investments, even when the stock market has gone through difficult times. On the other hand, we have the medical devices industry, also trying to provide medical technology products for the diagnosis and treatment of diseases and anomalies. The financial situation of such medical devices companies has fluctuated more than that of those in the pharmaceutical–biotechnology fields but for the most part they have held their own, even in turbulent financial times.

In the realm of medical devices, companies involved in the design, development, manufacturing, testing, and FDA approval of medical electronic devices are at the forefront of medical technology. Over the last 30 years, such technological advances as magnetic resonance imaging, nuclear radiation, optical diagnostical devices, EKG and EEG measurement devices, pacemakers, defibrillators, hearing aids, and portable diagnostic meters (glucose, oximetry, pulse analyzers) have made the diagnosis and treatment of medical ailments a much more productive endeavor, saving millions of lives and postponing death for many millions more.

The boundary between human and machine is disappearing. In science fiction and films such as those in the *Star Trek* series the boundary has almost completely disappeared. For example, in *Star Trek* episodes the sick bay of the starship *Enterprise* is full of gadgets that make the boundary between human and machine irrelevant. One of the most common devices used by the doctor is a "sensor," which is floated around a person's body and is capable of identifying the person's ailment. Another often used device is a syringe with no needles; the medication is directly absorbed by the skin and clothing does not seem to be a barrier. Although these medical advances may not happen in our lifetimes, it is possible that many of us will find ourselves wearing or carrying one type or another of electronic medical device at some point in our lives. Products derived from nanotechnology and micromechanical systems (MEMS) have already become a reality in many laboratories and research facilities, and one day soon may become the state of the art of the medical device market.

Today, many millions of people have some kind of medical device implanted in their bodies. Implantable electronics are now competing with or complementing pharmaceutical and other treatments for such ailments as tachycardia, Lou Gehrig's disease, Parkinson's disease, muscle spasticity, irregular breathing, diabetes, and deafness. Implantable electronic products include drug pumps, monitors, and delivery systems; cochlear implants; and neurostimulators. Among the nonimplantable electronic devices that have revolutionized medicine are diagnostic machines such as EKGs, EEGs, MRIs, PET scans, x-rays, CT scans, and optical devices and treatment machines such as, for example, radiation, particle accelerators, heart-lung bypass, and dialysis.

This book attempts to address the design of some of these electronic devices using an engineering approach but with sufficient introductory science. Not all existing devices can be addressed, but a sufficient number have been treated to provide the flavor of general design approaches. This material is most useful for engineers involved in the design of medical electronics, but it is also useful to those in the biomedical field with limited electronic background, since the approach is generally at the system level.

The book is divided into eight chapters. Power electronics is considered first because of the importance of proper power design for the overall performance of an electronic medical device. Considerable effort is spent addressing sensors in several chapters because sensors are the most important component of an electronic medical device. The art of data acquisition and digital signal processing is also considered extensively using a practical approach. In all of the discussions, examples of medical electronic devices are included.

1 PROPER DESIGN OF POWER SUBSYSTEMS IN MEDICAL ELECTRONICS

1.1. ELECTROMAGNETIC INTERFERENCE REQUIREMENTS

The safety and reliability of a switching-mode power supply (SMPS) for medical equipment depend basically on the design of power supplies with stringent requirements so as to minimize the possibility of failures. Therefore, it is important to balance as many necessary requirements as possible to achieve confidence that the medical device will perform according to the prescribed objectives. Failure to mitigate the potential for failure and satisfy such requirements in the early stages of design could completely negate the design approaches used in SMPS design.

1.2. TRANSIENT VOLTAGE PROTECTION

Many power supplies are exposed to stress conditions in the power bus in the form of transient voltages and currents that can reach 6000 V and 3000 A. The IEEE findings presented in Table 1.1 show the typical amplitudes and wave shapes at various locations that result from a variety of sources. The findings in this table are published in IEEE standard 587-1980 (or ANSI C6H.1-1980).

The IEEE standard shows that for indoor systems (ac lines less than 600 V), the transient waveform has an oscillatory shape as shown in Figure 1.1. These transients can excite natural resonance frequencies in the power bus system. Therefore, many of the transients become oscillatory in nature with different amplitudes and wave shapes at different locations in the power bus. According

TABLE 1.1 Surge voltages and currents deemed to represent the indoor environment and recommended for use in designing protective systems

IEEE Std. 587 location category	Comparable IEC 664 category	Waveform (impulse)	Medium exposure amplitude (impulse)	Type of specimen or load circuit	Energy (J) deposited in a suppressor with a clamping voltage of 500 V (120-V system)	Energy (J) deposited in a suppressor with a clamping voltage of 1000 V (240-V system)
A. Long-branch circuits and outlets	II	0.5 μs 100 kHz	6 kV	High impedance	—	—
			200 A	Low impedance	0.8	1.6
B. Major feeders, short-branch circuits, and load center	III	1.2/50 μs	6 kV	High impedance	—	—
		8/20 μs	3 kA	Low impedance	40	80
		0.5 μs	6 kV	High impedance	—	—
		100 kHz	500 A	Low impedance	2	4

TRANSIENT VOLTAGE PROTECTION

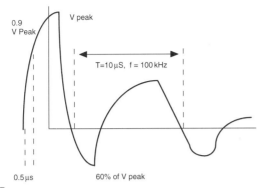

FIGURE 1.1 Proposed 0.5-μs, 100-kHz ring wave (open circuit voltage).

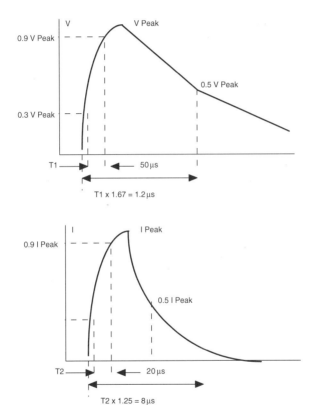

FIGURE 1.2 Unidirectional wave shapes (ANSI/IEEE Standard 28-1974).

to the IEEE standards, the frequencies of these surges and transients range from 5 to 500 kHz. For locations close to a service entrance (category B) much larger transients can be encountered, of the types shown in Figure 1.2.

The transient protection for these types of surges must be able to withstand the energies specified in Table 1.1. Furthermore, ring wave oscillatory conditions can also be observed and their voltage can reach 6 kV with a current of 500 A.

There is a variety of transient suppression devices. For the lower stress, category A locations, silicon varistors, transient suppression diodes, filter inductors, and capacitors are usually employed. Under the higher power of category B locations, these protection devices are supplemented with higher current rated gas discharge tubes on spark gaps.

Varistors display a voltage-dependent resistance characteristic. There is a turnover voltage in a metal-oxide varistor (MOV). At voltages below the turnover voltage, these MOVs have high resistance and low circuit loading. When the terminal voltage is greater that the turnover voltage, the resistance decreases rapidly and the surge current will flow in the shunt-connected varistor.

The MOV has the advantage of its high transient energy absorption capability, but its main disadvantage is its progressive degradation with repeated stresses and large slope resistance. Therefore, high slope resistance means that the MOV clamping action is quite poor for high current stress conditions, and as a result high voltages can be let through to the equipment to be protected. MOVs should be used in combination with other suppressor devices. In Figure 1.3 we see the MOV performance characteristics.

Transient protection diodes, which may be unidirectional or bidirectional, can also be used for general transient suppression. These are basically avalanche voltage clamp devices with high transient capability. In a bipolar protector, these junction diodes are placed in series "back to back."

Transient suppression diodes have the advantage of very high speed clamping actions (avalanche conditions). Another advantage is the low slope resistance in the conduction range. In the active region, the slope resistance is low, with the terminal voltage increasing by only a few volts at the transient currents of hundreds of amperes. Therefore, clamping diodes do a good job up to diode's maximum capabilities. Transient suppression diode performance characteristics are shown in Figure 1.3.

Transient suppressor diode performance characteristic line filters are one of the most effective methods for transient suppression. A typical line filter that can be found in category devices is shown in Figure 1.4. The inductors L_1 and L_2 and the capacitors C_1 through C_4 form the normal noise filter network. At the input to the filter network, varistors MOV1 through MOV3 provide the first level of protection from transients in the bus line. The inductances and varistors protect against small transients. For larger transients, the currents in L_1 and L_2 will eventually conduct charge to capacitors C_1 through C_4, reaching a voltage at which the suppressing diodes begin to prevent the output voltage from exceeding their rated clamp values, up to the current level at which the diodes will fail. This filter setup also protects the external power source from being affected by transients generated within the equipment power supply. This filter design protects against the differential mode (live to neutral) and against the common mode (live and neutral to ground), which although rare, is possible and very dangerous—hence the need for protection against it.

For category B transient suppression, the filter design of Figure 1.5 should be used. In the figure, we have an additional common mode inductor L_3. The inductors and capacitors C_1 through C_5 provide a good filter for common mode and series mode line conducted transients and radio frequency interference (RFI) noise. In addition the filter contains the varistors and a gas discharge

TRANSIENT VOLTAGE PROTECTION

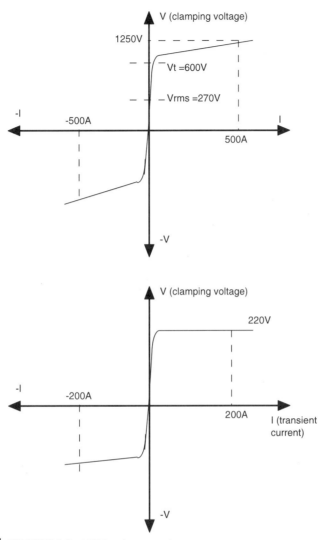

FIGURE 1.3 MOV performance characteristics.

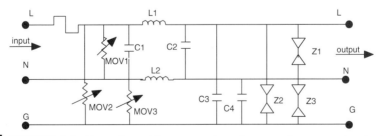

FIGURE 1.4 Line-to-line and line-to-ground transient overvoltage protection circuit with a noise filter, using MOV protection devices (low to medium power applications).

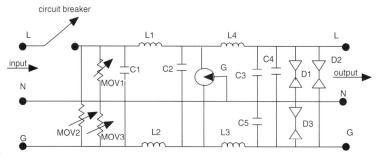

FIGURE 1.5 Line-to-line and line-to-ground transient protection circuit with a noise filter, using MOVs and transient protection diodes (for medium- to high-power applications).

arrester tube (G). In a gas discharge arrester, two or more electrodes are spaced within a sealed high-pressure inert gas environment. When the large voltage transient exceeds the voltage of the gas tube, an ionized glow discharge is first developed between the electrodes. As the current increases, an arc discharge is produced, providing a low-impedance path between the internal electrodes. The performance of the filter design in Figure 1.4 is similar to that of Figure 1.5 except that when a really large transient occurs, the gas arrester will reach the striking voltage, effectively short-circuiting all the lines and sending the transient to ground.

1.3. ELECTROMAGNETIC INTERFERENCE

Most switching mode power supplies (SMPSs) are capable of providing a considerable amount of conducted interference. The electromagnetic interference (EMI) is caused by the fast rectangular switching action, which is required for good efficiency.

1.3.1. Electromagnetic Interference Requirement

Most SMPSs are capable of providing a considerable amount of conducted interference. The EMI is caused by the fast rectangular switching action, which is necessary for good efficiency. Good design practices require the minimization of the RFI transmitted into the power supply itself or the output lines (from the power supply). Some of the RFI can also be radiated. Therefore federal and international regulations limit the amount of either conducted or radiated interference.

The power supply designer must study the regulations mandated by the United States and the European Common Market (the FCC and IEC regulations, respectively). Since June 1998, all medical devices have been required to comply with the Medical Devices Directive. With respect to electromagnetic compatibility (EMC), EN 60601-1-2: 93 is the standard. EN 60601-1-2 is identical to IEC 60601-1-2 and is the collateral standard to the IEC 60601-1, which is the general safety standard for medical electrical equipment. IEC 60601-1-2, the IEC 60601 standard was first published in April 1993 (Medical Electrical

ELECTROMAGNETIC INTERFERENCE

TABLE 1.2 The proposed interference tests for medical equipment

IEC 60601 (to be updated in mid 2001)

Emissions EN 55011 privilege class B (only for special units class A); EN 55014 for click noise

Immunity *ESD* (IEC 61000-4-2): 2 + 4 + 6 kV contact, 2 + 4 + 8 kV air.

Radiated Field (IEC 61000-4-3): 3 V/m, 80 to 2500 MHz, AM: 80%, 1 KHz or 2 Hz, modulation frequency; 10 V/m from 80 to 2500 MHz for life-supporting systems.

Burst (IEC 61000-4-4): 2 kV, all kinds of power supplies; 1-kV input/output lines >3 m; patient cables are excluded.

Surge (IEC 61000-4-5): 0.5- and 1-kV differential mode; 0.5, 1, and 2 kV common mode.

Conducted RF Immunity (IEC 61000-4-6): 3 V, 80% AM, 0.15 to 80 MHz, 1-kHz or 2-Hz modulation frequency; 10 V for life-supporting frequencies at the ISM frequencies.

Magnetic Fields (IEC 61000-4-8): 50 to 60 Hz, 10 A/m; special units can have less.

ac Variations (IEC 61000-4-11): For systems with input power less than 1 kW, voltage test levels 0, 40, and 70% duration (periods) 0.5, 5, 25; voltage interruption 5 s.

Harmonics (IEC 61000-3-2/3/4/5): TBD.

Failure Criteria: Clinical utility is maintained; performance criteria and special tables with information for the user must be included.

Equipment: Part 1: General Requirements for Safety; Part 2: Collateral Standard: Electromagnetic Compatibility — Requirements and Test). The collateral standard EN 60601-1-2:93 contains the emissions and immunity requirements. The emissions tests are specified in EN 55011 Group I, II, Class A,B. The immunity requirements are identified in EN61000-4-3 and EN 61000-4-5. The tests described in EN 60601-1-2:93 are well known to most medical device manufacturers and are shown in Table 1.2.

Of the two types of interference, conducted interference is the most significant since such conducted interference can be a major (if not the greatest) contributor. Techniques to decrease conducted EMI will likewise decrease radiated EMI. The two major aspects of conducted interference are differential mode interference and common mode interference. Differential mode conducted interference is the RF noise that exists between any two supply or output lines. In the case of SMPSs, this would be the live and neutral ac supply lines or the positive and negative lines in the dc supply lines. The interference voltage is shown in Figure 1.6.

There are really several sources of both conducted and radiated interference. An illustration of this is shown in Figure 1.7. The failure to screen the switching devices (i.e., switching field-effect transistors [FETs]) and failure to RF screen the transformer are principal causes of conducted common mode interference. Common mode interference is also the most difficult to eliminate using filters because of the limited decoupling capacitor size.

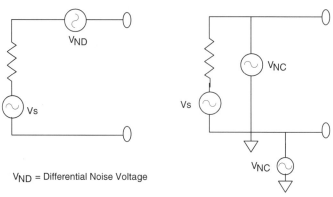

V_{ND} = Differential Noise Voltage

V_{NC} = Common Mode Noise Voltage

FIGURE 1.6 Representation of differential and common mode noise.

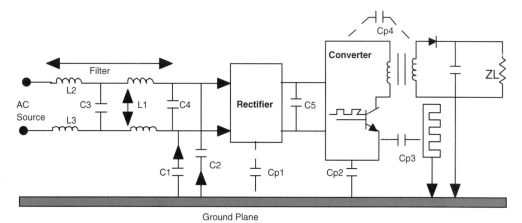

FIGURE 1.7 Sources of conducted and radiated interference.

In Figure 1.7, the differential noise is easily suppressed by the storage capacitors and the large decoupling capacitors C_3 and C_4, which are located across the supply lines.

Common mode RF current is introduced into the local ground by insulation leakage and parasitic capacitive coupling, as shown by Cp1 through Cp4 in Figure 1.7. The return path of these noise currents will be near the input supply lines through decoupling capacitors C_1 and C_2. These noise currents become a constant current source to capacitors C_1 and C_2 because the source voltage and source impedance are very high. The noise voltage across C_1 and C_2 is

$$V_{nh} = I_n \times \frac{1}{j\omega C_{1(2)}}$$

where

$\dot{V}_{nh}$ = harmonic interference voltage
I_n = interference noise current at the harmonics frequency
ω = harmonic frequency.

The voltage source V_{nb} will drive the noise current into series inductors L_1, L_2, and L_3 and into the output lines to return via the ground plane. This external RF current causes EMI, and hence it is covered in EMI regulations for the purpose of suppression and reduction of R_d current in the ground plane of the source as the best method for EMC compliance. Once the common mode current gets into the ground plane, it is very difficult to eliminate it and even more difficult to predict the path it will take. All high-voltage ac components should be isolated from the ground plane. Transformers should have faraday cages, which should be returned to the input dc lines to divert capacitively coupled common mode current to the supply lines. The RFI cages are used in addition to the safety screen that must be returned to the ground plane for safety reasons.

In Figure 1.7, inductor L_1 is known as the common mode choke. It must have a high common mode inductance and also carry the 60-Hz supply current. To accomplish both tasks, a high-permeability core material should be used. It is normal to wind L_1 with two windings. The windings carry a long current at twice the frequency, as the rectifier diodes conduct only at the peak of the input voltage waveforms. The windings are built such that they provide maximum inductance for common mode currents but cancel for series mode currents. This phasing prevents the core from becoming saturated at 60 Hz. Common mode noise shows up on both supply lines of L_1 with respect to ground. The large shunt capacitor C_2 ensures that the noise amplitude is the same on both lines where they connect to the inductor. The two windings will behave as one since they will be in phase, hence providing one large common impedance. To maintain a good high-frequency rejection, the self-resonance frequency of the choke must be as high as possible. Therefore, the parasitic capacitance of the coil must be kept as low as possible, which is the reason why single layer windings are used on insulated high-permeability ferrite toroids.

The inductance provided by a common mode choke is given by

$$L = N^2 A_L \tag{1.1}$$

where A_L is the inductance factor (in nanohenries) for a given ferrite toroid and N is the number of turns for the choke. The numbers of turns N is given by

$$N = \sqrt{\frac{R_{cu}}{A_r}} \tag{1.2}$$

Here R_{cu} is the dc resistance of the copper wire used for winding and A_r is the copper resistance factor for the chosen ferrite toroid. It was previously stated that faraday cages may be needed to shield several components in SMPSs that would decrease or eliminate parasitic capacitances. The easy way to eliminate undesirable coupling is to place electrostatic screens between the components and the heat sink. The screen, normally copper, must be insulated from both the heat sink and the switching transistor or diode so that it capacitively couples the ac current and returns to a single-point ground in the input circuit. Examples of faraday cages are shown in Figure 1.8.

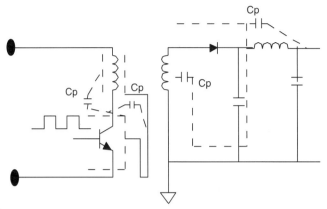

FIGURE 1.8 Faraday cage protection of components.

1.4. INRUSH CURRENT CONTROL

In direct-off-line switch mode power supplies, the input power is directly fed from the ac power lines without using the large 60-Hz isolation transformer normally found in linear power supplies. In the switching mode power supply, the input-to-output isolation is provided by a smaller high-frequency transformer. To provide a dc input to the converter, a rectification is first performed using semiconductor power rectifiers and large electrolytic capacitors. The correct size of the rectifiers, diodes, input fuses, and filter is very important. To size these components correctly, a full knowledge of the relevant applied stress is required.

In direct-off-line switch mode supplies, it is common practice to use direct-off-line semiconductor bridge rectification with a capacitive input filter to produce a high-voltage dc supply for the converter section. When the line input is switched directly to this type of rectifier capacitor arrangement, very large inrush currents will flow. This is highly stressful to the components and will cause interference when other equipment share a common supply line impedance. The basic rectifier capacitor input filter and energy storage circuit is shown in Figure 1.9. The effective series resistance R_s is made up of all the various series components, including the source resistance, that appear between the prime power source and reservoir capacitors C_1 and C_2. To further reduce peak currents, additional series resistance may be added to provide a final optimum effectiveness series resistance. The series resistor must be such that it would withstand initial high-voltage and high-current stress. High-current surge-rated resistors are best suited for this application, mostly wire wound types.

A modified inrush circuit in combination with a powerful fast-acting start system is shown in Figure 1.10. The transistor $Q1$ is a high-voltage transistor. In this circuit, R_1 and R_2 are not only limiting current resistors but are used for biasing transistor $Q1$ into a fully saturated region after the initial switch on of the supply.

SOFT START

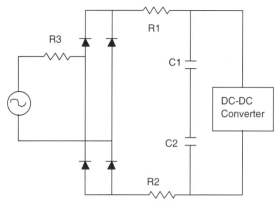

FIGURE 1.9 Rectifier capacitor input filter.

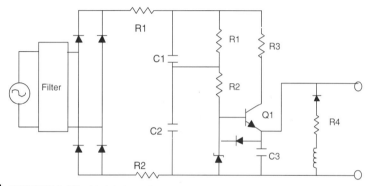

FIGURE 1.10 Modified rectifier capacitor input filter.

1.5. SOFT START

Soft start is different from inrush limiting. Soft start acts on the converter control circuit to give a progressively increasing pulse width. This progressive start not only reduces the inrush current stress on the output capacitors and converter components, but it also reduces the problem of transformer flux doubling. It is normal for SMPSs to take the line input directly to the rectifier and a large storage filter capacitor via a low-impedance noise filter. In order for the input capacitor to fully charge during start-up, it is necessary to delay the start-up of the power converter so that it will not draw current from the input capacitors until they are fully charged. To address this principle, a start-up delay and soft start procedure are usually provided by the control circuit. This will delay the initial switch-on of the converter and allow the input capacitors to fully charge. In Figure 1.11, we see the soft start circuit added to the transistor start circuit.

However, even under this controlled turn-on condition, it is possible for the output voltage to overshoot as a result of race conditions in the control circuit. The overshoot can be considerably reduced by making the soft start action slow. A circuit suitable for decreasing the overshoot is shown in Figure 1.12.

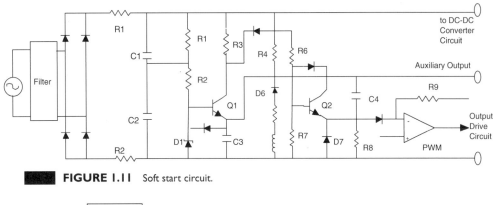

FIGURE 1.11 Soft start circuit.

FIGURE 1.12 Overshoot decreasing circuit.

1.6. OVERVOLTAGE PROTECTION

In case of fault conditions, most power supplies have the potential to deliver higher output voltages than those normally required. To protect the equipment under these conditions, it is common practice to provide some means of overvoltage protection within the power supply.

In low-power applications, overvoltage protection may be provided by a simple clamp action. In many cases, just a shunt Zener diode is all that is needed to provide the required overvoltage protection (see Figure 1.13a). If a higher current capability is required, a more powerful transistor shunt regulator can be used (Figure 1.13b).

Another form of voltage limiting is shown in Figure 1.14. In the circuit, an optocoupler is energized in the event of a voltage condition. A small-signal SCR triggers on the primary circuit to switch off the primary converter.

1.7. UNDERVOLTAGE PROTECTION

In most power systems, a sudden rapid increase in load current results in dips in power supply line voltages. This is caused by the rapid increase in current during

UNDERVOLTAGE PROTECTION

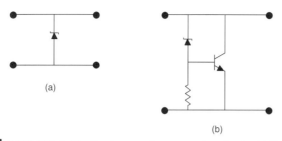

FIGURE 1.13 Simple overvoltage protection circuits. (a) Shunt Zener diode, (b) shunt regulator.

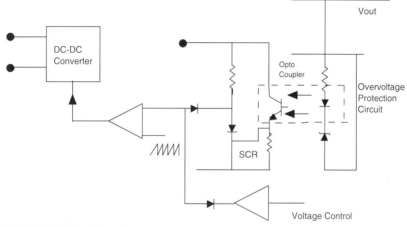

FIGURE 1.14 Voltage-limiting circuit with an optocoupler.

the transient demand and the limited response time of the power supply. Even in cases when the transient performance of the power supply is excellent, the voltage at the load can still dip when the load is located at large distances and the inductance and resistances of the lines become significant. If the load variations are relatively small, it is sufficient to provide a low-impedance capacitor at the load or of the supply lines to "hold up" the voltage during transient loading. If the loads are large and hence load variations are also large, large shunt capacitors are required.

The undervoltage protection circuit should be located as close to the load as possible, as shown in Figure 1.15.

An example of an undervoltage circuit is shown in Figure 1.16. When Q_1, which acts as a switch, opens capacitors C_1 and C_2 are charge in parallel from the supply lines in parallel via resistors R_1 and R_2. These capacitors will charge to the supply voltage V_s. In the figure, the circuit in its charged state is connected to a linear regulator circuit at the input. If during an undervoltage condition Q_1 closes and conducts, capacitors C_1 and C_2 are connected in series and provide a voltage of $2V_s$ at the terminals. Because the voltage at the input of the linear regulator now exceeds the required output voltage V, Q_1 can operate as a linear regulator, supplying the needed transient current and maintaining the output voltage across the load nearly constant. This will continue until C_1 and C_2 have discharged to half their initial voltage.

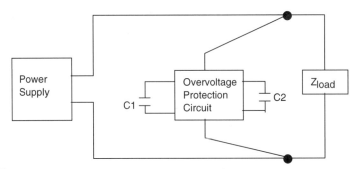

FIGURE 1.15 Position voltage protection circuits close to the power supply.

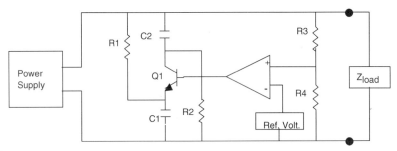

FIGURE 1.16 Undervoltage detection circuit.

1.8. OVERLOAD PROTECTION

Most power supplies must provide full overload protection. This includes short circuit protection and current limits on all outputs. The purpose of overload protection is to protect both the power supply and the load. Loads are usually protected by specifying a requirement that the load current should not be greater than 20% of the specified current rating for the load.

A very effective method for overload protection is based on foldback output current limiting. Not only the output voltage of the power supply but also the output current is decreased as the load resistance moves to zero as the result of a fault.

In a purely resistive load, the load line will be a straight line, as shown in Figure 1.17a. As a resistive load changes, the straight line will swing clockwise around the origin to become horizontal for a short circuit (zero resistance). It can be observed that a straight resistive load line can cross the foldback characteristic of the power supply at only one point (e.g., $P1$ in Figure 1.17a). When the maximum limiting current I_{max} has been reached at $P2$, any further increase of the load (i.e., reduction of load resistance) results in a reduction in both output voltage and current.

In nonlinear loads, however, instability (or "lockout") can occur and a smooth shutdown will not occur unless the power supply is brought out of lockout mode. Nonlinear loads are found in most semiconductor circuits. This nonlinear behavior can be observed in Figure 1.17b.

SNUBBER CIRCUITS

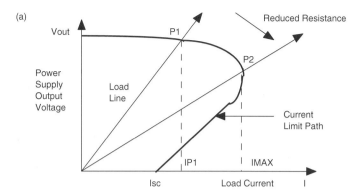

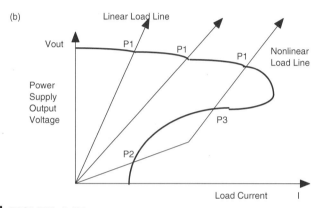

FIGURE 1.17 Load current for linear and nonlinear loads. (a) Current overload shutdown characteristic, (b) overload for linear and nonlinear load lines.

In nonlinear behavior, the load line crosses the power supply foldback current at three points. Points $P1$ and $P2$ are both stable operating points of the power supply. When the supply load ensemble is first turned on, the output voltage is only partially established to point $P2$, and lockout occurs. This happens because the slope resistance of the load line at point $P2$ is less than the slope of the power supply foldback characteristic at the same point. Since $P2$ is a stable point, lockout is maintained and the power supply is never turned on.

1.9. SNUBBER CIRCUITS

Snubber circuits are usually a dissipative combination of a resistor–capacitor–diode network and are filled across high-voltage switching transistors to reduce the switching stress and EMI problems during the on–off functions of the switching transistor. In bipolar transistors, the snubber circuit is required to give the load line shaping and to ensure that secondary breakdown, reverse bias, and safe operating limits are not exceeded. In off-line flyback converters, this is important since the flyback voltage can exceed 800 V.

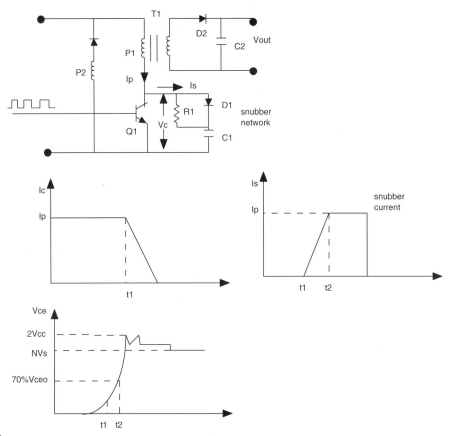

FIGURE 1.18 An illustration of a snubber circuit and associated signal representations.

Figure 1.18 shows a single-ended flyback converter. The snubber circuit, composed of components $D1$, $C1$, and $R1$, is placed across the collector to the emitter of $Q1$. In Figure 1.18, we also see the voltage and current waveforms from this circuit. One of the main functions of the snubber circuit is to provide an alternative path for the inductive current I_p as Q turns off. It is then possible to turn off $Q1$ without a significant rise in its collector voltage during the turn-off. Without the snubber circuit, the voltage on $Q1$ would be very large, defined by $V = L(dI/dt)$. Since the snubber reduces the rate of change of the collector voltage during the turnoff edge, it reduces RFI problems.

In this example, $C1$ is chosen such that the voltage on the collector V_{ce} is 70% of the V_{ceo} rating of $Q1$ when the collector current has dropped to zero at time t_2. During the collector current fall time (t_1 to t_2), the current in $C1(I_s)$ will be increasing linearly from zero to I_p. The near current over the period will be $I_p/2$. The value of the optimum snubber capacitor $C1$ can be determined from

$$\frac{dV_c}{dt} = \frac{1}{2}\frac{I_p}{C1} \tag{1.3}$$

Hence, if the collector voltage is to be more than 70% of V_{ceo} when the collector current reaches zero at time t_2, then

$$C1 = \frac{I_p[t_2 - t_1]}{2(70\% V_{ceo})} \tag{1.4}$$

where I_p is the maximum primary current (in amps); $[t_2-t_1]$ is given in microseconds; the term V_{ceo} is the V_{ceo} rating for that transistor (in volts); and C1 is the snubber capacitance in microfarads. The power dissipated by the transistor $Q1$ during the turnoff period $t_2 - t_1$ is given by

$$P_{Q1}(\text{off}) = \tfrac{1}{2} C1 (70\% V_{ceo})^2 f \quad \text{in mW} \tag{1.5}$$

where

$C1$ = the snubber capacitance in microfarads
f = frequency in kilohertz
$V_{ceo} = V_{ce}$ rating of the transistor (70% V_{ce} is the chosen maximum voltage at $I_c = 0$)

The snubber resistor value is given by

$$R1 = \frac{1}{2} \frac{t_{\text{off}}(\min)}{C1} \tag{1.6}$$

In the process of measuring the turnoff current, the designer should consider the effects of the Miller current that will flow into the collector capacitance during turnoff. It results in an apparent collector current conduction, even when $Q1$ is turned off. The magnitude depends on the rate of change of collector voltage dV_c/dt and the collector to base depletion capacitance. If there is a sizable heatsink for $Q1$, there may be considerable capacitance between the collector of $Q1$ and common heat sink, providing an additional path for the collector current. Maximum collector dV_c/dt values are sometimes given in the data sheets of switching transistors (BJTs or FETs), and this can be satisfied by a suitable selection of C1 in the snubber circuit.

1.10. OUTPUT FILTERING

High-frequency radiated and conducted ripple noise must be controlled. Faraday screens can be used in transformers and between high-frequency, high-voltage components and the ground plane. To reduce conducted noise at the output, filtering may well also be required.

Normally, to provide a steady dc output with very little ripple noise, LC low-pass filtering is used. In forward converters, these types of filters carry out two main functions. The first is to serve as energy storage to maintain a steady dc output voltage throughout the power-switching cycle. The second objective is to reduce high-frequency conducted series and common mode output interference. However, to maintain a nearly constant dc output voltage, the current in the output capacitor must also be nearly constant; hence a considerable inductance is required in the output inductor. The inductor, which is relatively large,

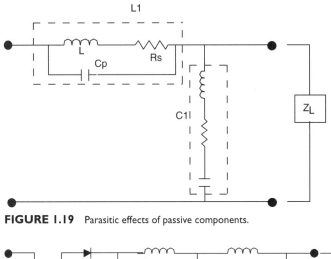

FIGURE 1.19 Parasitic effects of passive components.

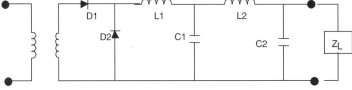

FIGURE 1.20 Second-stage filter.

can have parasitic capacitance, giving a relatively low self-resonance frequency. These inductors will have a low impedance at frequencies above self-resonance and will not give good attenuation of the higher-frequency components of the conducted interference. Likewise, the output capacitor will have parasitic resistance and inductance, causing the output capacitor high-frequency attenuation to be very poor. These parasitic effects can be seen in Figure 1.19.

A more cost-effective wideband filter can be obtained by using a second stage; much smaller inductance and capacitance values are required in the second stage. In Figure 1.20, the first capacitor C1 is selected to be quite large. The first inductor $L1$ is designed to carry the maximum load current with minimum load and without saturation. Suitable core materials such as permalloy or iron-dust toroids should be used. The second inductor $L2$ must have the maximum impedance at high frequency, and requires a low interwinding capacitance. This provides a high self-resonant frequency. Also $L2$ can be built from a small ferrite bobbin, and small iron-dust toroids. Normal ferrite material can be used for a ferrite rod inductor, and large air gaps prevent dc saturation. The second capacitor is also much smaller than C1. It is chosen for low impedance at high frequencies. Capacitor C2 consists of a small electrolytic capacitor shunted by a low-inductance ceramic capacitor. Since $L1$ and $L2$ conduct large dc currents, the term "choke" has been widely used in the past.

The filter just described will not be effective for common mode noise. The common mode noise voltages appear between the output lines and the ground plane. The common mode noise is caused by parasitic inductive and capacitive coupling between the power circuits and the ground plane. The best approach

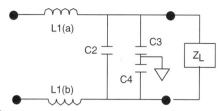

FIGURE 1.21 Filter for common mode noise.

for minimizing common mode noise is to provide the best layout possible for the power supply. Further reduction of the common mode noise can be obtained by splitting inductors $L1$ and $L2$ into two parts to form a balance filter, as shown in Figure 1.21. Additional capacitances $C3$ and $C4$ are required between each output line and the ground plane to provide a return path for the common mode noise current.

1.11. POWER FAILURE WARNING

Protecting medical devices from the consequences of a power failure by the use of power failure warning circuits is an important consideration. Warning concerning an imminent power failure is necessary to provide sufficient time for an organized system shutdown. One of the most normal processes in a power failure is first a fall in the inline voltage to a value below the normal minimum, a process called partial brownout. The brownout condition can eventually cause a permanent failure. An example of a power failure detection circuit is shown in Figure 1.22.

A small portion of the dc voltage on the power converter capacitors $C1$ and $C2$ is compared with a reference voltage by the comparator of $A1$. If this voltage falls to a value at which the power supply would provide the prescribed hold time, then the output of amplifier $A1$ goes to a high output, the optical coupler is excited, and a failure warning is generated. This requires that the power supply provides sufficient holdup time from a minimum defined input voltage to ensure that the specified warning period is satisfied before the output voltage fails. The selected warning voltage must be lower than the minimum dc voltage normally found on $C1$ and $C2$ under fully loaded, minimum-line-input voltage conditions. To provide the needed holdup time in a fully loaded condition, from this lower capacitor voltage, the dc–dc converter will give full output for a supply voltage that is even lower than normal; hence larger capacitors are required and larger current rated input components are required. This makes the power supply larger and more expensive.

It is possible to detect the imminent failure of the line before it has fully developed by observing the rectified line input. The circuit can respond to the reduction in the dv/dt, which occurs at the beginning of a half-cycle of operation if the peak voltage is going to be low. Therefore, the power supply is able to give an earlier warning of impending low-voltage conditions. If the dv/dt is low as the supply passes through zero, failure is assumed, and a warning signal

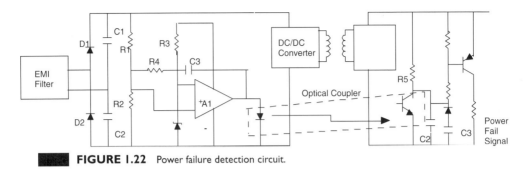

FIGURE 1.22 Power failure detection circuit.

is generated before the half-cycle is complete. This can provide a few more milliseconds of warning. The scenario is shown in Figure 1.23.

1.12. FLIGHTBACK SWITCH MODE POWER SUPPLIES

In flightback converters, secondary regulators are often of linear dissipative types; however, switching regulators can be used to obtain a higher efficiency. The closed loop of the main output can minimize the dissipation. The closed-loop control regulation can also be shared among several outputs. The main attraction of flightback converters is their simplicity and low cost.

There are basically two modes of operation that are clearly identifiable in flightback converters: (1) the complete energy transfer in which all the energy that was stored in the transformer during an energy storage period is transferred, and (2) the incomplete energy transfer in which a part of the energy stored in the transformer during the "on" period remains in the transformer at the beginning of the next "on" period. A flyback power rectifier converter is shown in Figure 1.24.

When the transistor $Q1$ is turned on, the start of the winding in the transformer $T1$ goes positive. The output rectifier diode $D1$ will go to reversed bias and will not conduct. There will be no current in the secondary $S1$ while $Q1$ conducts. In the energy storage phase, only the primary winding is active, and the transformer in the primary behaves as a series inductor. The current through the primary increases at the rate

$$\frac{dI_P}{dt} = \frac{V_{cc}}{L_p} \qquad (1.7)$$

where V_{cc} is the supply voltage and L_p is the primary inductance. The flux density of the transformer increases from B_1 to B_2, as shown in Figure 1.25.

When $Q1$ turns off, the primary current drops to zero. The transformer cannot change without a change in the flux density ΔB. The change in flux density goes negative and the voltage will reverse on all windings (flies back). The secondary rectifier diode $D1$ will conduct, and the magnetizing current will now transfer to the secondary. It will continue to flow in the secondary winding. The secondary current flows in the same direction in the secondary winding as

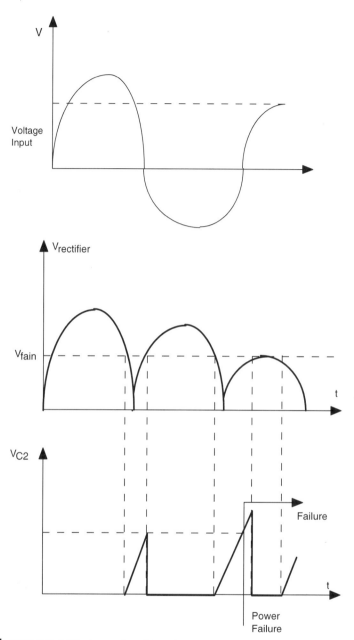

FIGURE 1.23 Low-voltage failure detection.

the original current flowed in the primary winding, but the magnitude is defined by the turns ratio.

In a steady state condition, the induced electromagnetic force (emf) (flyback voltage) must have a value in excess of the voltage in C1 before diode $D1$ can conduct. The secondary current is given by $I_S = nI_p$ (n is the transformer turn ratio). The secondary current decays at a rate specified by the secondary voltage

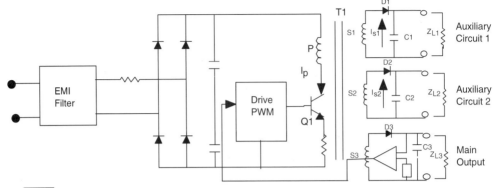

FIGURE 1.24 Overview of the flyback converter.

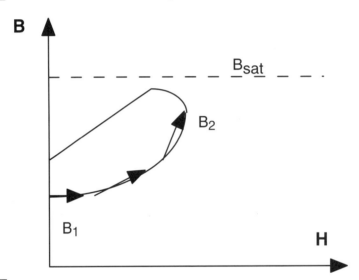

FIGURE 1.25 Flux density changes in a flyback converter.

V_{S1} and the secondary inductance L_{S1}.

$$\frac{dI_{S1}}{dt} = \frac{V_{S1}}{L_{S1}} \qquad (1.8)$$

where V_{S1} is the secondary voltage of the auxiliary circuit.

1.13. HALF-BRIDGE FLYBACK CONVERTER

The half-bridge converter is also known as the two-transistor FET converter. This type of circuit is shown in Figure 1.26. The high-voltage dc line is switched to the primary of the transformer by two power FET transistors FET1 and FET2. These switches are driven by the control circuit such that both FETs are either on or off. The flyback action takes place during the off state. A small drive transformer is used to provide simultaneous drive signals to the two FET

HALF-BRIDGE FLYBACK CONVERTER

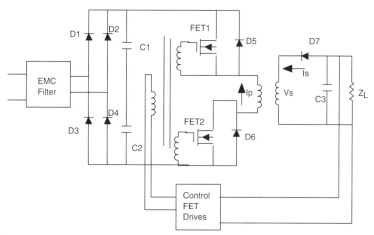

FIGURE 1.26 Half-bridge flyback converter.

switches. The cross-connected diodes $D5$ and $D6$ return excess flyback current to the supply lines and provide voltage clamping of FET1 and FET2 at a value of only one diode drop above and below the supply line voltage.

The energy recovery action of diodes $D1$ and $D2$ eliminates the need for an energy recovery winding or large snubber elements. The voltage and current waveforms are shown in Figure 1.27.

When FET1 and FET2 are on, the supply voltage will be applied across the transformer primary L_p. The starts of all windings will go positive and the output rectifier diode $D7$ will be reverse biased and on–off. This means that secondary current will not flow during the "on" period and the secondary leakage inductance can be neglected. During the "on" period, current increases linearly in the transformer primary defined by the equation

$$\frac{dI_P}{dt} = \frac{V_{cc}}{L_p} \qquad (1.9)$$

The energy $E = \frac{1}{2} L_p I_p^2$ is stored in the coupled magnetic field of the transformer. At the end of the "on" period, transistors FET1 and FET2 will turn off simultaneously and the primary supply current in the FET will fall to zero. The magnetic field strength cannot change without a corresponding change in the flux density, and by flyback action, all voltage on the transformer will reverse. Diodes $D6$ and $D5$ are brought into conduction, clamping the primary flyback voltage to the supply voltage. Because the polarity is reversed on all windings, the secondary voltage V_s will also bring the output rectifier diode $D7$ into conduction. The current is increased in the secondary winding. When the secondary current reaches a value of $n \times I_p$, where n is the turns ratio, the energy recovery clamp diodes $D1$ and $D2$ cease conduction and the primary voltage V_p falls back to the reflected secondary voltage. The voltage across the primary will be the voltage across C3. The clamped flyback voltage is less than the supply voltage V_c; otherwise the flyback energy will be returned to the supply voltage. In a normal process and in a complete energy transfer, the remaining energy

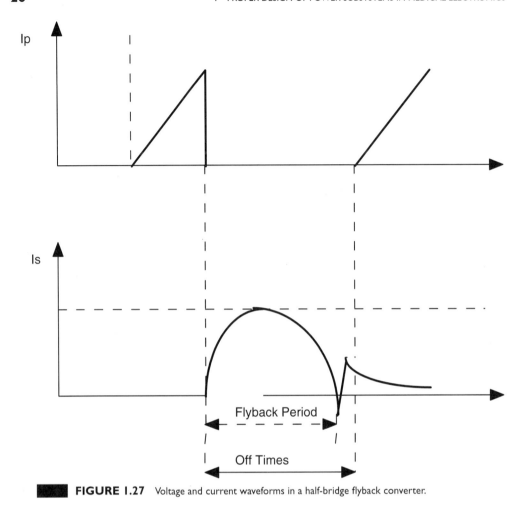

FIGURE 1.27 Voltage and current waveforms in a half-bridge flyback converter.

stored in the transformer magnetic field is transformed to the output capacitor and the load while FET1 and FET2 are in the "off" period. A new power cycle is initiated after the off period.

1.14. FORWARD CONVERTER

A singled-ended forward converter is shown in Figure 1.28, where the output inductor L_s carries a large dc current component. In the forward converter, the secondary windings $S1$ are phased such that the energy will be transferred to the output circuit when the transistor power is "on." The power transformer $T1$ operates as a true transformer with a low output resistance, and the inductance L_s is required to limit the current flow in the output rectifier $D1$, the output capacitor (out and the load).

When the transistor $Q1$ turns on, the supply voltage V_s is applied to the primary winding $P1$. A secondary voltage V_s is developed and applied to the

FORWARD CONVERTER

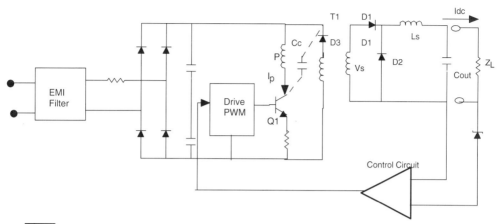

FIGURE 1.28 Schematic overview of the forward converter.

rectifier $D1$ and inductance L_s. The voltage across L_s is about $V_s - V_{out}$. The current in L_s is increasing linearly and is defined by the equation

$$\frac{dI}{dt} = \frac{V_s - V_{out}}{L_s} \tag{1.10}$$

At the end of the "on" period, $Q1$ will turn off and the secondary voltages will reverse by the normal flyback action of $T1$. Current $T2$ continues to flow in the forward current under the effects of the inductance L_s, and diode $D2$ conducts. The voltage across the inductance L_s is reversed with a value equal to the output voltage (if we neglect diode voltage drops). The current in L_s decreases as defined by the equation

$$\frac{-dI}{dt} = \frac{V_{out}}{L_s} \tag{1.11}$$

The voltage applied to L_s in the forward and reverse directions must be equal for steady state conditions; therefore, when the on and off periods are equal, the output voltage will be $V_s/2$. The inductor current is the required output I_{dc}. When the ratio of the "on" time to the "off" time decreases from the 50% duty factor, the output voltage will fall until forward and reverse voltage equality is obtained. The output is defined by the following equation

$$V_{out} = \frac{V_s \times t_{on}}{t_{on} + t_{off}} \tag{1.12}$$

where V_s is the secondary voltage (in volts), t_{on} is the time that $Q1$ is conducting in microseconds, and t_{off} is the time that $Q1$ is "off" in microseconds. The ratio $t_{on}/(t_{on} + t_{off})$ is called the duty ratio.

There are several advantages of the forward converter when compared to the flyback converter.

1. The copper losses in the forward converter transformer are somewhat lower because the peak currents in primary and secondary will tend to

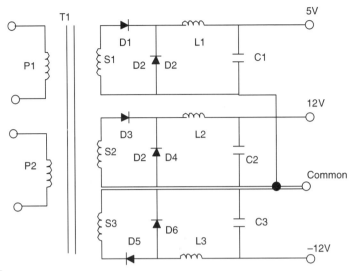

FIGURE 1.29 Multiple-output forward converter.

be lower than in the flyback case (the inductance is higher). This results in a smaller temperature increase in the transformer.
2. The reduction in secondary ripple current is sustained. The action of the output inductor and flywheel diode maintains a constant current in the output load. Because the energy stored in the output inductor is available to the load, the output capacitor can be made small with its main function being to reduce output ripple voltages. The ripple current rating for this capacitor is much lower than that required by the flyback case.
3. Output ripple voltages are lower.

The main disadvantages of the forward converter are as follows:

1. Higher cost is incurred due to the extra output inductor and flywheel diode.
2. In light load situations, when L_s reverts to the discontinuous mode, excessive output voltage will be produced unless minimum loads are specified.

In Figure 1.29 we observe a typical multiple-output forward converter secondary in which all outputs share a common return line. The negative output is developed by reversing $D5$ and $D6$. We also notice that the phasing of the secondary is such that $D3$ and $D5$ conduct at the same time during the "on" period of $Q1$.

1.15. HIGH-VOLTAGE DEFIBRILLATORS

Devices such as pacemakers and defibrillators are miniature processors that use sensitive low-voltage, low-power, application-specific integrated circuits

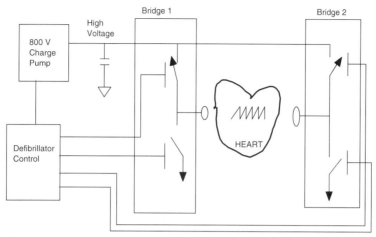

FIGURE 1.30 Block diagram of defibrillator.

(ASICs) to monitor, regulate, and control the delivery of electrical pulses to the heart. Implantable cardiovascular defibrillators (ICDs) have been in use for several years. When such devices detect a life-threatening cardiac fibrillation, the ICD applies a high-voltage pulse between two electrodes connected to the heart. The pulse can be as high as 800 V, with the resulting current reaching tens of amperes (during a few milliseconds). The high voltage is generated and stored on a large capacitor through the usage of a charge pump. Normally, the shock is delivered to the heart via a two-phase pulse. Figure 1.30 shows a principal block diagram of a two-phase defibrillator system that features a typical high-voltage bridge required to generate the biphasic pulse.

The specific application shown in Figure 1.30 contains two identical bridges, each having two switches: one switch connected to ground and the other to the high voltage. Insulated-gate bipolar transistors (IGBTs) are used as the switch elements. These transistors offer the minimum on-resistance (R_{on}) relative to the silicon area. The high side of these transistors requires a gate voltage that is from 10 to 15 V higher than the voltage to be switched. A transformer is used for level shifting between the high-voltage controller and the switch. Figure 1.31 shows a block diagram of the components needed for one of the bridges. The use of a transformer can cause some problems because such transformers are bulky and difficult to manufacture.

To overcome the limitations of transformers, a design such as shown in Figure 1.32 can be contemplated. With a low-voltage interface, the bridge can be controlled directly using a complementary metal-oxide semiconductor (CMOS) level controller IC with complete integration of all the necessary components for the bridge. Some of the main features include a small form factor using a ball-grid array (BGA) multichip module package (MCM). The BGA approach allows for the complete circuits to be packaged at a considerable size reduction. To obtain the smallest possible form factor and quiescent current, the bridge design calls for high-side level shifting and charge transfer via capacitive

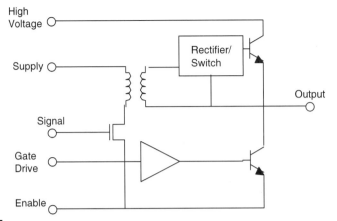

FIGURE 1.31 Bridge with transformer as isolation for the high-side driver.

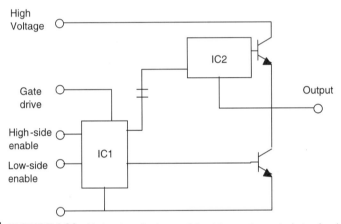

FIGURE 1.32 High-voltage bridge module with capacitor as isolation for the high-side driver.

coupling achieved by integrated circuits $IC1$ and $IC2$ in Figure 1.32. Low-gate-charge IGBTs enable the use of lower capacitance on the isolation–charge transfer capacitor.

REFERENCES

Cuk, Slobodan. 1983. "Analysis of integrated magnetics to eliminate current ripple in switching converters." *PCI Conference Proceeding*, April.

Dowell, P. L. 1966. "Effects of eddy currents in transformer windings." *Proc. IEE* 113(8).

IEEE. IEEE Std. 1980. 587-1980. "IEEE guide for surge voltages in low voltage AC power circuits." ANSI/IEEE C62-41-1980.

Janson, L. 1973. "A survey of converter circuits for switched mode power supplies." *Mullard Tech. Commun.* 12 (119).

Billings, Keith. 1999. *Switching Power Supply Handbook*. McGraw Hill, New York.

Middlebrook, R. D. 1976. "Input filter considerations in design and application of switching regulators." *IEEE Industrial Applications Society Annual Meeeting Record*, October.

REFERENCES

Middlebrook, R. D. 1978. "Design techniques for preventing input filter oscillation in switched mode regulator." *Proc. Powercon.* 5, May.

Pearson, W. R. 1982. "Designing optimum snubber circuit for the transistor bridge configuration." *Proc. Powercon.* 9.

Royer, G. H. 1955. "A switching transistor dc to AC converter having an output frequency proportional to the DC input voltage." *AIEE*, July.

Snigier, Paul. 1980. "Those sneaky switchers." *Electronic Products*, March.

Snigier, Paul. 1981. "Power supply selection criteria." *Electronic Design*, August.

Switchers Pursue Linear Below 199 W. 1981 Electronics Products, September.

Venable, D. H., and S. R. Foster. 1982. "Practical techniques for analyzing, measuring and stabilizing feedback control loops in switching regulators and converters." *Proc. Powercon.* 7.

2
FUNDAMENTALS OF MAGNETIC RESONANCE IMAGING

2.1. EARLY HISTORY OF NUCLEAR MAGNETIC RESONANCE

The root of nuclear magnetic resonance (NMR), or as it sometimes is called, magnetic resonance imaging (MRI), goes back to World War II. In 1938 I. Rabi perfected a beam-splitting technique and successfully achieved nuclear magnetic resonance—a term that Rabi coined. The NMR experiments made use of the spin-state–dependent force that inhomogeneous magnetic fields exert on an atomic beam of silver atoms directed perpendicular to the gradient fields. For spin-$\frac{1}{2}$ nuclei, the atomic beam splits, but it reconverges when the polarity of the gradient field reverses. Rabi showed that irradiating the spins at the transition frequency, which interchanges the $m = \pm\frac{1}{2}$ states, eliminated the convergence.

The first detection of NMR in bulk matter was achieved in the mid-1940s by research groups led by Edward M. Purcell at Harvard University and Felix Bloch of Stanford University. Purcell used a resonant cavity to study the absorption of radio-frequency energy in paraffin: at resonance, the cavity output was found to be slightly reduced. By contrast, Bloch and his colleagues used what they called "nuclear induction." Bloch described the experiment as measuring an electromotive force resulting from the forced precession of the nuclear magnetization in the applied RF field.

When matter is placed in a magnetic field, the nuclear magnetic moments orient parallel to the field, leading to a paramagnetic polarization in the direction of the magnetic field—the z direction. If an oscillating magnetic field is applied in the x or y direction, the polarization vector is deflected from the z direction once the field approaches the resonance value. The resonance condition is given by

$$|\gamma|B = \omega$$

where B is the amplitude of the applied static magnetic field, ω is the nuclear precession frequency, and γ is the gyromagnetic ratio, which is a constant for a given isotope. In NMR, this rotation of the magnetic polarization vector of the nuclei in a plane perpendicular to the z axis induces an emf in a detector coil; this is the NMR signal.

Nuclear magnetic relaxation—that is, the return of the spin system to equilibrium is of great significance to imaging and was conceptualized by these early investigators. By repeatedly passing the spin system through the resonance condition (by varying the amplitude of the polarizing magnetic field) and observing the reappearance of a signal, Bloch found that for protons (that is, the hydrogen nuclei) in liquids the time constant T_1 for the return of the longitudinal magnetization was of the order of seconds. Further, he concluded from the sharpness of the resonance that the spins' phase memory time—the transverse relaxation time T_2—is of the order of hundreds of milliseconds in fluid (and, as shown later, only slightly shorter in biological tissues). It is clearly thanks to Mother Nature's good graces, or God, that NMR in human subjects is possible at all. If it took, instead of seconds, hours for the spin to repolarize, the technique would be impractical.

The commonly used detection method during the first two decades of NMR work exploited the principles of continuous wave excitation, where the field is swept while the sample is irradiated with RF energy of constant frequency. An alternative scheme, which is still in use, consists of pulsed RF excitation followed by the detection of the resultant preprecession signal. Hence, rather than being simultaneous, in this scheme excitation and detection are performed sequentially.

A major milestone was the discovery of the chemical shift by Warren Proctor, F. C. Yu, and W. C. Dickinson. They found that in ammonium nitrate, two nitrogen-14 resonances could be observed, which they ascribe to the different chemical environments to which the nitrogen nucleus is exposed in the nitrate and ammonium ions. Similar findings were later made by others for nuclei such as fluorine, phosphorous, and hydrogen. These observations constitute the basis of modern NMR spectroscopy. A few years later, as the magnetic homogeneity that determines the frequency resolution achievable in NMR was improved further, another type of fine structure was discovered. This structure, which is due to spin–spin coupling, is fundamental to modern high-resolution spectroscopy, and together with the chemical shift provides the basic ingredients for molecular structure determination. Today, NMR is the preeminent method for determining the structures of biomolecules with molecular masses up to 100,000 daltons.

The development of pulse Fourier transform by Richard Ernst and Weston Anderson was of great importance for NMR. This alternative mode of signal creation, detection, and processing led to an unprecedented enhancement in per-unit-time detection sensitivity compared with continuous wave excitation techniques. If N channels are used simultaneously in an experiment, then, provided the dominant source of noise is not the excitation, the sensitivity increases by a factor of $N^{1/2}$. Ernst and Anderson demonstrated that one can affect broadband excitation by exciting the nuclear spins with short RF pulses of a single carrier frequency.

A new breakthrough was added to NMR technology in 1973 when Paul Lauterbur at the State University of New York at Stony Brook first proposed generating spatial maps of spin distributions by what he called "NMR zeugmatography." The key to this method was the idea of superimposing magnetic field gradients onto the main magnetic field to make the resonance frequency a function of the spatial origin of the signal. In the presence of a magnetic field gradient the frequency domain signal is the equivalent of a projection of the object onto the gradient axis. By rotating the gradient in small angular increments, one obtains a series of projections from which an image can be reconstructed using back projection techniques.

In and around 1980 whole body experimental NMR scanners were in operation, and by 1981 clinicians began to explore the clinical potential of magnetic resonance imaging. NMR has several advantages over x-ray computerized tomography (CT). First, it was noninvasive—that is, it did not require ionizing radiation or the injection of contrast material. Second, it provided intrinsic contrast far superior to that of x-ray CT. Some of this early work showed that magnetic resonance imaging was uniquely sensitive to diseases of the white matter of the brain, such as multiple sclerosis. Further, the contrast could be controlled to a significant extent by the nature and timing relationships of the radio-frequency pulses. Third, MRI was truly multiplanar and even three dimensional—that is, it could provide images in other than the traditional transverse plane without the subject having to be repositioned. This property turned out to be of great value over x-ray CT for the study of the brain and other organs.

In 1975 Ernst introduced a new class of NMR experiments now known as two-dimensional NMR, and should be regarded as the parent of modern NMR techniques. One can understand the principle by reference to Figure 2.1. Suppose the spins in a point object of spin density $\rho(x, y)$ are excited by an RF pulse in the presence of a magnetic field gradient of magnitude G_y, which we call a phase-encoding gradient. These spins will resonate at a relative frequency $\omega = \gamma B(y)$, where $B(y) = G_y Y$. If the gradient is active for a period t_y, the phase at the end of the gradient period is $\phi y = \gamma G_y y t_y$. Let us then step the time t_y in equal increments, as applied by Figure 2.1b. We readily notice that the phase at the end of the gradient period varies cyclically with time. At time $t = t_y$, the gradient G_y is turned off and an orthogonal gradient G_x is applied for a duration t_x, during which the free-induction decay signal is collected. During the detection period the spins precess at a frequency $\omega x = \gamma G_x x$. One thus encodes spatial information into both the phase and the frequency of the NMR signal, whose magnitude is proportional to $\rho(x, y)$. Of course, in the case of imaging a real object, the signal has a multitude of frequency and phase components.

An obvious drawback of stepping the duration of the gradient G_y is the decrease in signal amplitude due to the irreversible decay (with relaxation time T_2) of the transverse magnetization. Because the phase shift imparted to the signal by the gradient G_y is a function of the gradient's duration as well as its amplitude, one can achieve the same effect by stepping the amplitude of the

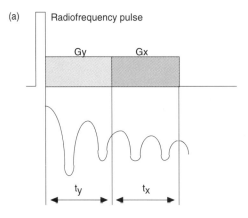

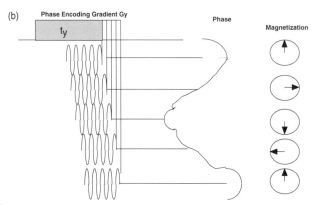

FIGURE 2.1 Fourier zeugmatography principle.

gradient while keeping its duration constant. This important modification of the technique gave rise to what is called spin-warp imaging. The term refers to the "warping" of the phase caused by the gradient. Raising the amplitude of the first gradient incrementally during each excitation and read cycle yields an $N_1 \times N_2$ array of raw data, from which one can reconstruct an image by double Fourier transformation of the signal. The resulting digital image consists of $N_1 \times N_2$ picture elements, or pixels, whose values are proportional to the amplitudes of the detected transverse magnetizations. The process of incrementally increasing a gradient in amplitude to encode spatial information into phases of processing spins is called phase encoding.

The basis of the signal and contrast in NMR is the transient nature of the signal. Spins, following excitation, return to their equilibrium state with characteristic time constants, the spin relaxation times T_1 and T_2, which for water in biological tissues are of the order of hundreds of milliseconds or even seconds. This process can be described by the phenomenological Block equations, which predict the evolution of the spin system in terms of the longitudinal and transverse components of the complex spin magnetization.

Consider a typical imaging experiment in which radio-frequency pulses are applied repeatedly at time intervals $\tau < T_1$ for the purpose of spatial encoding. Then the magnetization available for detection is attenuated by a factor $1 - e^{-\tau/T_1}$. This, in itself, would not be a sufficient mechanism for modulating the image signal were it not for the large range of relaxation times found for the water protons in mammalian tissues—from about 100 ms to several seconds.

Though the process of tissue water relaxation is not completely understood, its rate is related to the extent of binding of water to the surface of biological macromolecules. Increased binding slows molecular motion. The more closely the reorientation motion of the magnetic dipoles matches the Larmor frequency, the greater the transition probabilities between the nuclear energy levels, and thus the greater the relaxation rates. A case in point is the brain, the majority of which consists of gray and white matter adjacent to fluid-filled cavities, the ventricles. Because water is more tightly bound in white matter than in gray matter, the water molecules in white matter reorient more slowly than those in gray matter, thus more closely matching the Larmor frequency; hence $T_1\text{wm}$ (wm is for white matter) is less than $T_1\text{gm}$ (gm is for gray matter). By contrast, spinal fluid, which from the point of view of molecular motion closely parallels neat water, has much faster molecular motion, and thus $T_1\text{sf}$ is much greater than both $T_1\text{wm}$ and $T_1\text{gm}$.

The equilibrium magnetization is proportional to the proton concentration in the tissues, hence the different plateau values. For short pulse-recycle times ($\tau \ll T_1$) the signal amplitudes follow the reciprocal of t_1, whereas at long recycle times ($\tau \gg T_1$) they are governed by their equilibrium magnetization—that is, they follow the proton concentrations. Contrast in MRI is therefore a continuum, and unlike in x-ray imaging, there is no universal gray scale. Further, it was recognized early on that in most diseased tissues, such as tumors, the relaxation times are prolonged. This difference provides the basis for image contrast between normal and pathological tissues.

2.2. GENERAL REVIEW OF MRI

Protons (hydrogen nuclei) precess when placed in a magnetic field. This phenomenon is the basis for nuclear magnetic resonance. Nuclear precession occurs with a frequency directly proportional to the strength of the magnetic field, with a proportionality constant called the gyromagnetic ratio, of about 42.6 MHz per tesla. Typical frequencies range from 300 to 800 MHz (see Figure 2.2).

The precessional axis lies along the direction of the magnetic field. If an oscillating magnetic field at the precessional frequency is applied perpendicular to the static field, the protons will now precess about the axis of the oscillating field, as well as that of the static field. The condition is known as nutation. The oscillating field is generated by a tuned RF resonator, or RF coil, which usually surrounds the sample. The magnetic field of the precessing protons induces, in turn, an oscillating voltage in the RF coil, which is detected when the RF field is gated off. This voltage is then amplified and demodulated to baseband, as in a normal superheterodyne receiver, and digitized using an analog-to-digital converter (see Figure 2.3).

38 2 FUNDAMENTALS OF MAGNETIC RESONANCE IMAGING

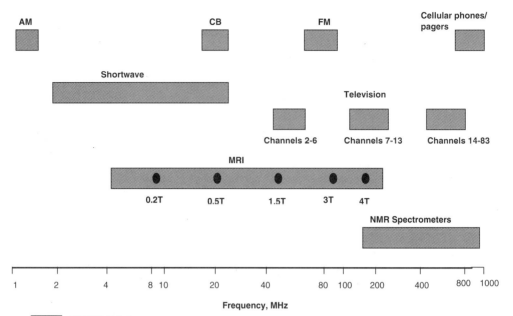

FIGURE 2.2 Magnetic resonance frequencies.

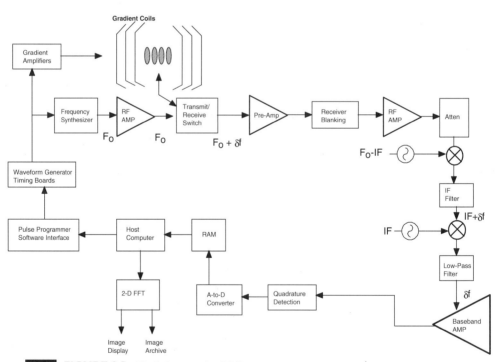

FIGURE 2.3 Block diagram of an NMR system.

Little information of practical use would attach to the demodulated signal if all the protons exhibit identical precessional frequencies. In fact, very minor perturbations arise in the proton precessional frequency that reflect the molecular environment of the protons. In an ethanol molecule (CH_3CH_2OH), for example, the protons in the CH_3 group precess at a slightly different frequency from those in the CH_2 group or the OH group. Spectroscopy can determine structural information from the Fourier spectrum of the received signal.

To factor out the signal's dependence on the static magnetic field, NMR measurements are often given in a unitless quantity called chemical shift, which is typically measured in parts per million. It is the difference between the precession frequency of protons that are part of a particular molecular group and that of protons in a reference compound, that makes NMR possible to be measured. This application of magnetic resonance is generally referred to as high-resolution NMR spectroscopy, and is widely used in the pharmaceutical and chemical industries.

NMR imaging, better known as magnetic resonance imaging, is one of the most important imaging technologies found in most modern hospitals. In MRI, it is the protons in the water molecules of a patient's tissue that are the source of the signal. The spatial information needed to form images from magnetic resonance is obtained by placing magnetic field gradient coils on the inside of the magnet. These coils, constructed from copper wire, create additional magnetic fields that vary in strength as a linear function of distance along the three spatial axes (Figure 2.4). This means that the resonant frequencies of the water protons within the patient's body are now spatially encoded.

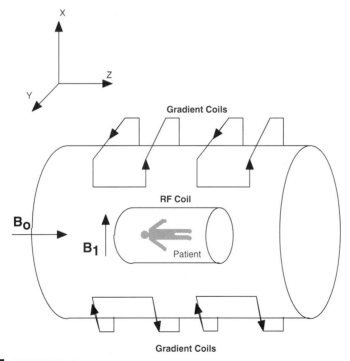

FIGURE 2.4 RF and gradient coils construction.

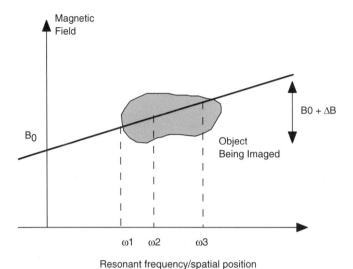
FIGURE 2.5 Spatial position caused by magnetic field.

The chief use of clinical MRI is for imaging the brain and spine. However, the recent development of rapid imaging methods has extended its application to the chest and abdomen where motion previously caused blurring of the image. The contrast in MRI images arises from differences in the number of protons in a given volume and in their relaxation times (the time taken for the magnetization of a sample to return to equilibrium after the RF pulse is turned off), which are related to the molecular environment of the protons (see Figure 2.5). Following excitation, each proton within the excited volume precesses at the same frequency. During detection of the echo, a gradient B_0 is applied causing a variation in the frequencies for the protons generating the echo signal. The frequency of precession ω_i for each proton depends upon its position. Frequencies measured from the echo are mapped to the corresponding position.

Brain scans are carried out to detect tumors, infarcts, aneurysms, or other pathological conditions. These are normally easy to detect since both the proton densities and relaxation times are markedly different from those of healthy tissue.

2.3. A MORE DETAILED OVERVIEW OF MRI

In clinical practice, magnetic resonance imaging has been a great tool in the medical community. Over the years, MRI has assisted physicians in diagnosis, treatment, and presurgical treatments. The main advantage of MRI when compared with other imaging tools (e.g., x-ray computed tomography or CT scan) is that it does not require exposure of the human body to ionizing radiation; therefore it is very safe. The MRI signals are also very sensitive to different parts of tissues.

2.3.1. The Physics of Spin

Spin is a fundamental property of nature like electrical charge or mass. Spin comes in multiples of $\frac{1}{2}$ and can be $+$ or $-$. Protons, electrons, and neutrons all possess spin. Individual unpaired electrons, protons, and neutrons each possesses a spin of $\frac{1}{2}$. In the deuterium atom (^2H), with one unpaired electron, one unpaired proton, and one unpaired neutron, the total electronic spin $= \frac{1}{2}$ and the total nuclear spin is equal to 1. Two or more particles with spins having opposite signs can pair up to eliminate the observable manifestations of spin. An example is helium. In nuclear magnetic resonance, it is unpaired nuclear spins that are of importance.

Spin Properties

When placed in a magnetic field of strength B_0, a particle with a net spin can absorb a photon of frequency υ. The frequency depends on the gyromagnetic ratio γ of the particle.

$$\upsilon = \gamma B_0 \tag{2.1}$$

For hydrogen, $\gamma = 42.58 \, \text{MHz/T}$.

Almost every element in the periodic table has an isotope with a nonzero nuclear spin. NMR can be performed only on isotopes whose natural abundance is high enough to be detected. However, some of the nuclei that are of interest in MRI are listed in Table 2.1.

To understand how particles with spin behave in a magnetic field, consider a proton. This proton has the property called spin. Think of the spin of this proton as a magnetic moment vector, causing the proton to behave like a tiny magnet with north and south poles. When the proton is placed in an external magnetic field, the spin vector of the particle aligns itself with the external field, just like a magnet would. There is a low-energy configuration of state where the poles are aligned N-S-N-S and a high-energy state N-N-S-S.

Transitions

A particle can undergo a transition between the energy states by the absorption of a photon. A particle in the lower energy state absorbs a photon and ends up

TABLE 2.1 Net spin of several nuclei of interest for MRI

Nuclei	Unpaired protons	Unpaired neutrons	Net spin	γ (MHz/T)
^1H	1	0	$\frac{1}{2}$	42.58
^2H	1	1	1	6.54
^3P	0	1	$\frac{1}{2}$	17.25
^{23}Na	2	1	$\frac{3}{2}$	11.27
^{14}N	1	1	1	3.08
^{13}C	0	1	$\frac{1}{2}$	10.71
^{19}F	0	1	$\frac{1}{2}$	40.08

in the upper energy state. The energy of this photon must exactly match the energy difference between the two states. The energy E of a photon is related to its frequency υ by Planck's constant $(h = 6.62 \times 10^{-34} \, \text{J s})$.

$$E = h\upsilon \qquad (2.2)$$

In NMR and MRI, the quantity υ is called the resonance frequency and the Larmor frequency, respectively.

The energy of the two spin states can be represented by an energy level diagram. We have seen that $\upsilon = \gamma B_0$ and $E = h\upsilon$; therefore the energy of the photon needed to cause a transition between the two spin states is

$$E = h\gamma B_0 \qquad (2.3)$$

When the energy of the photon matches the energy difference between the two spin states an absorption of energy occurs. In the NMR, the frequency of the photon is in the radio-frequency (RF) range. In NMR spectroscopy, υ is between 60 and 800 MHz for hydrogen nuclei. In clinical MRI, υ is typically between 15 and 80 MHz for hydrogen imaging.

When a group of spins is placed in a magnetic field, each spin aligns in one of the two possible orientations. At room temperature, the number of spins in a lower energy level, N^+, slightly outnumbers the number in the upper level, N^-. Boltzmann statistics tells us that

$$\frac{N^-}{N^+} = e^{-E/kT} \qquad (2.4)$$

where E is the energy difference between the spin states; k is Boltzmann's constant, 1.3805×10^{-23} J/K; and T is the temperature in kelvin. As the temperature decreases, so does the ratio N^-/N^+. As the temperature increases, the ratio approaches one.

The signal in NMR spectroscopy results from the difference between the energy absorbed by the spins that make a transition from the lower energy state to the higher energy and the energy emitted by the spins that simultaneously make a transition from the higher energy state to the lower energy state. The signal is therefore proportional to the difference between the states. NMR is a rather sensitive spectroscopy since it is capable of detecting these very small perturbation differences. It is the resonance, or exchange of energy at a specific frequency between the spins and the spectrometer, that gives NMR its sensitivity.

It is worth noting two other factors that influence the MRI signal: the natural abundance of the isotope and biological abundance. The natural abundance of an isotope is the fraction of nuclei having a given number of protons and neutrons, or atomic weight. For example, there are three isotopes of hydrogen: ^{1}H, ^{2}H, and ^{3}H. The natural abundance of ^{1}H is 99.985%. Table 2.2 lists the natural abundances of some nuclei studied by MRI. The biological abundance is the fraction of one type of atom in the human body. Table 2.3 lists the biological abundance of some nuclei studied by MRI.

TABLE 2.2 Natural abundance of isotopes of interest to MRI

Element	Symbol	Natural abundance
Hydrogen	^{1}H	99.985
Hydrogen	^{2}H	0.015
Carbon	^{13}C	1.11
Nitrogen	^{14}N	99.63
Nitrogen	^{15}N	0.37
Sodium	^{23}Na	100
Phosphorus	^{31}P	100
Potassium	^{39}K	93.1
Calcium	^{43}Ca	0.145

TABLE 2.3 Biological elements of interest to MRI

Element	Biological abundance
Hydrogen (H)	0.63
Sodium (Na)	0.00041
Phosphorus (P)	0.0024
Carbon (C)	0.094
Oxygen (O)	0.26
Calcium (Ca)	0.0022
Nitrogen (N)	0.015

Spin Packets

It is cumbersome to describe NMR on a microscopic scale. A microscope picture is more convenient. The first step in developing the microscopic picture is to define the spin packet. A spin packet is a group of spins experiencing the same magnetic field strength. In this example, the spins within each grid section represent a spin packet.

At any instant in time, the magnetic field due to the spins in each spin packet can be represented by a magnetization vector. The size of each vector is proportional to $(N^{+} - N^{-})$. The vector sum of the magnetization vectors from all of the spin packets is the net magnetization. To describe pulsed NMR, it is necessary to talk in terms of the net magnetization. Adapting the conventional NMR coordinate system, the external magnetic field and the net magnetization vector at equilibrium are both along the z axis.

T_1 Processes

At equilibrium, the net magnetization vector lies along the direction of the applied magnetic field B_0 and is called the equilibrium magnetization M_z which equals M_0. We refer to M_z as the longitudinal magnetization. There is no transverse (M_x or M_y) magnetization here. It is possible to change the net magnetization by exposing the energy of a frequency equal to the energy difference between the spin states. If enough energy is put into the system, it is possible to saturate the spin system and make $M_z = 0$. The time constant that describes how M_z returns to its equilibrium value is called the spin lattice relaxation time (T_1). The equation governing this behavior as a function of the time t after its displacement is

$$M_z = M_0 \left(1 - e^{-t/T_1}\right) \tag{2.5}$$

Therefore T_1 is defined as the time required to change the Z component of magnetization by a factor of e.

If the net magnetization is placed along the $-Z$ axis, it will gradually return to its equilibrium position along the $+Z$ axis at a rate governed by T_1. The equation governing this behavior as a function of the time t after its displacement is

$$M_z = M_0 \left(1 - 2e^{-t/T_1}\right) \tag{2.6}$$

The spin-lattice relaxation time (T_1) is the time to reduce the difference between the longitudinal magnetization (M_z) and its equilibrium value by a factor of e.

If the net magnetization is placed in the XY plane it will rotate about the Z axis at a frequency equal to the frequency of the photon that would cause a transition between the two energy levels of the spin. This frequency is called the Larmor frequency.

T_2 Processes

In addition to the rotation, the net magnetization starts to dephase because each of the spin packets making it up experiences a slightly different magnetic field and rotates at its own Larmor frequency. The longer the elapsed time, the greater the phase difference. Here the net magnetization vector is initially along $+Y$. For this and all dephasing examples you can think of this vector as the overlap of several thinner vectors from the individual spin packets.

The time constant that describes the return to equilibrium of the transverse magnetization M_{xy} is called the spin–spin relaxation time T_2.

$$M_{XY} = M_{XY_0} e^{-t/T_2} \tag{2.7}$$

T_2 is always less than or equal to T_1. The net magnetization in the XY plane goes to zero and then the longitudinal magnetization grows in until we have M_0 along the z axis.

Any transverse magnetization behaves the same way. The transverse component rotates about the direction of applied magnetization and dephases. Time T_1 governs the rate of recovery of the longitudinal magnetization.

In summary, the spin–spin relaxation time T_2 is the time to reduce the transverse magnetization by a factor of e. In the previous sequence, T_2 and T_1 processes are shown separately for clarity. That is, the magnetization vectors are shown filling the XY plane completely before growing back up along the

Z axis. Actually, both processes occur simultaneously with the only restriction being that T_2 is less than or equal to T_1.

Two factors contribute to the decay of transverse magnetization:

1. molecular interactions (said to lead to a pure T_2 molecular effect)
2. variations in B_0 (said to lead to an inhomogeneous T_2 effect)

The combination of these two factors is what actually results in the decay of transverse magnetization. The combined time constant is called T_2^*. The relationship between the T_2 from molecular processes and that from inhomogeneities in the magnetic field is as follows.

$$\frac{1}{T_2^*} = \frac{1}{T_2} + \frac{1}{T_2 \text{ inhomo}}. \tag{2.8}$$

Magnetic resonance imaging is based on the physical principle of nuclear magnetic resonance. Nuclear magnetic resonance was discovered by Black and Purcell in 1946. When certain atomic nuclei are placed in a static magnetic field, it will assume either a higher energy level or a lower energy level (Figure 2.6). As shown in the figure, the energy between the two states is linearly dependent on the strength of the applied static field: a physical principle known as the Zeeman effect. A nucleus in a higher energy state can fall to the lower energy state. A nucleus will emit a photon when transitioning between these two energy states. The energy of the photon is equal to the energy difference between the two states. A nucleus in the lower energy state can jump to a higher energy state, absorbing a photon with energy matching the energy difference between the two states. This means that when nuclei under the effect of an applied magnetic field are irradiated by photons in the form of electromagnetic fields at a certain frequency by an RF probe, some nuclei in the lower energy states will absorb the photons and jump to a higher energy state. After the radiated energy is interrupted, the nuclei in the higher energy state will return to the lower energy state to recover to equilibrium, emitting photons on electromagnetic

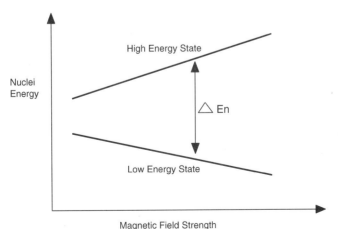

FIGURE 2.6 Energy level of nuclei under a magnetic field.

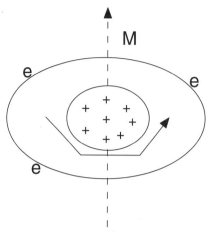
FIGURE 2.7 Magnetic field precession in magnetic fields.

fields, which can be detected by an MRI probe. The frequency emitted by the electromagnetic fields is detected by the energy difference between the two states of the nuclei. The decay of the signals in time depends on the molecular environment of the nuclei. Because the energy difference between the two states of a given nucleus in an external field depends on the strength of the external field, the energy difference at any point in the body can be made different by varying the magnetic field strength from point to point. Therefore, the energy of the photons and the frequency of the electromagnetic fields absorbed and emitted by the nuclei are also different from point to point. When the signals from all nuclei are received, their frequency can be used to determine spatial information about the nuclei.

From quantum mechanics certain atomic nuclei possess what is called spin. A proton, which has a mass, can be thought of as spinning, as shown in Figure 2.7 with its own angular momentum. The circulating positive current creates a small magnetic field.

Neutrons can also be thought of as generating a small magnetic field since their distributed positive and negative charges are not uniformly distributed. These small magnetic fields are called magnetic moments M and are given by

$$M = \gamma J \qquad (2.9)$$

where J is the angular momentum and γ is known as the geomagnetic ratio. For nuclei with an odd number of protons or an odd number of neutrons, there is never a net angular momentum equal to zero because of opposite spin states (Pauli exclusion principle that the angular momentum of each proton in a nucleus must assume opposite spin states). Many biological nuclei usually have odd numbers of protons and spins; example of such are ^1H, ^{13}C, ^{19}F, ^{23}N, and ^{31}P. Usually ^1H is the most significant nucleus in most MRI imaging because it is part of the water molecule and it has the highest NMR sensitivity with a geomagnetic ratio $\gamma/2\pi = 42.58$. Let's consider a magnetic flux B_0 in the z direction.

A MORE DETAILED OVERVIEW OF MRI

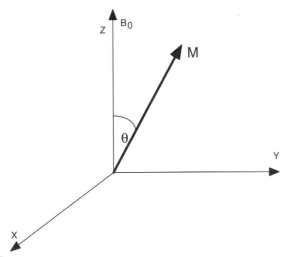

FIGURE 2.8 The magnetization vector.

The proton can assume two positions: with its z component of the magnetic moment aligned with the external field or with its z component of the magnetic moment opposed to the external field. Both states are stable, but the energy associated with the latter is higher. This can be observed in Figure 2.8 and the angle θ_0 can be determined by the value of the magnetic moment M and its z component M_z, which are given by quantum mechanics as

$$M = \frac{\gamma h \sqrt{3}}{4\pi} \qquad M_z = \frac{\gamma h}{4\pi} \qquad (2.10)$$

where h is the Planck constant ($h = 6.629 \times 10^{-34}$ J s). Therefore

$$\theta_0 = \cos^{-1}\left(\frac{M_z}{M}\right) = \cos^{-1}\left(\frac{1}{\sqrt{3}}\right) \approx 54.7° \qquad (2.11)$$

The difference in energy between the two states

$$\Delta E = 2M_z B_0 = h\upsilon \qquad (2.12)$$

Therefore the frequency υ is given by

$$\upsilon = \left[\frac{2M_z}{h}\right] B_0 \qquad (2.13)$$

It can be observed that the frequency is directly proportional to the magnetic field.

The angular momentum of a nucleus is linearly related to its magnetic moment

$$\frac{d\vec{M}}{dt} = \gamma \left(\vec{M} \times \vec{B_0}\right) \qquad (2.14)$$

Representing this equation in terms of scalar components and combining such equations we obtain

$$M(t) = a_x(M_x \cos(\gamma B_0)t + M_y \sin(\gamma B_0)) \\ + a_y(M_y \cos(\gamma B_0)t - M_x \sin(\gamma B_0)) + a_z M_z$$

This last expression of the magnetic moment has a frequency component given by

$$f = \frac{\upsilon B_0}{2\pi} \qquad (2.15)$$

which is known as the Larmor resonant frequency. It can be easily shown that the frequency of the radiation is the same as the precessional frequency of the magnetic moment.

When a magnetic field is applied to a bulk mass, each individual magnetic moment must align itself either with or against the external field. Alignment with the magnetic field involves a lower energy and is called the parallel state (α). Alignment against the magnetic field involves a higher energy and it is called the antiparallel state (β).

Boltzmann's law is given by

$$\frac{P_\alpha}{P_\beta} = \exp\left(\frac{\Delta E}{KT}\right) \qquad (2.16)$$

where P_α denotes the probability that a nucleus is found in the parallel state (α), P_β denotes the probability that a nucleus is found in the antiparallel state (β), K is Boltzmann's constant ($K = 1.3800 \times 10^{-23}$ J/K), T is the absolute temperature of the sample. For protons at 20°C, ΔE is on the order of 10^{-26} J and KT is on the order of 10^{-21} J. Therefore the previous equation can be expressed as

$$P_\alpha - P_\beta \approx \frac{\Delta E}{2KT} \qquad (2.17)$$

If we estimate, $P_\alpha \cong P_\beta \cong \frac{1}{2}$. The net magnetic moment per unit volume, known as magnetization, is given by the expression

$$M = (P_\alpha - P_\beta)nM_z\hat{a}_z \approx \frac{\Delta E}{2KT}nM_z\hat{a}_z \qquad (2.18)$$

where n denotes the number of protons per unit volume. Notice that as the temperature increases the magnetization is destroyed. Since ΔE is linearly dependent on B_0, the net magnetization is proportional to the magnetic field. In NMR the observation of the precession of this magnetic moment is of great importance.

The effects of RF radiation on the magnetization of the body sample in a uniformly applied B_0 field needs to be addressed. Under the influence of an RF magnetic field, directed on a given processes coordinate, the magnetization can be rotated away from its equilibrium position and the angle θ is given by

$$\theta = VB_{\text{eff}}T \qquad (2.19)$$

where B_{eff} is the value of the effective magnetic field in the rotating frame. The rotation of M will have an angular frequency

$$\omega = \gamma B_{eff} \quad (2.20)$$

We can notice that a large gyromagnetic ratio provides a quicker perturbation of the magnetization vector. In practice it rotates M by 90 and 180°. The durations of these rotations are given by

$$T_{90} = \frac{\pi}{2\gamma B_{eff}} \quad (2.21a)$$

$$T_{180} = \frac{M}{\gamma B_{eff}} \quad (2.21b)$$

After the initial 90° rotation, the magnetization vector M processes in the transverse plane. The relaxation process is better explained by the Block equations described next.

$$\begin{aligned}\frac{dM_{x,y}}{dt} &= \gamma(\vec{M} \times \vec{B})_{x,y} - \frac{M_{x,y}}{T_2} \\ \frac{dM_z}{dt} &= \gamma(\vec{M} \times \vec{B})_z + \frac{M_0 - M_z}{T_1}\end{aligned} \quad (2.22)$$

where T_2 and T_1 are the transverse and longitudinal relaxation times, respectively, and M_0 denotes the equilibrium value of magnetization, which is assumed to lie in the z direction. The solution of Eqs. (2.22) is given by

$$\begin{aligned}M_x(t) &= M_0 \exp\left(-\frac{t}{T_2}\right) \cos(\gamma B_0 t) \\ M_y(t) &= -M_0 \exp\left(-\frac{t}{T_2}\right) \sin(\gamma B_0 t) \\ M_z(t) &= M_0 \left[1 - \exp\left(-\frac{t}{T_1}\right)\right]\end{aligned} \quad (2.23)$$

The relaxation signal as it returns to equilibrium is shown in Figure 2.9.

When a magnetic field B_0 is applied in the z direction, the nuclei process about the z axis at the Larmor frequency and with a precessional angle θ. Some of the nuclei process around the $+z$ axis protons and others precess around the $-z$ direction and this results in a bulk magnetization, as shown in Figure 2.10.

Upon the application of an RF pulse phase coherence is accomplished in both the $+z$ and $-z$ precession. The RF pulse is at the resonant frequency and it stimulates the flipping between the two sections in the $+z$ and $-z$ precession. This allows energy to be imparted into the protons and nuclei migrate to the $-z$ precession. Once the population of nuclei in each $+z$ and $-z$ precession sections is the same, there will be no net axial magnetization. After a 90° pulse, as the magnetization processes in the transverse plane, two types of relaxation occur. The "longitudinal" relaxation causes the axial magnetization to be brought

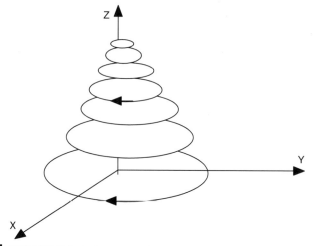

FIGURE 2.9 Relaxation signal and equilibrium.

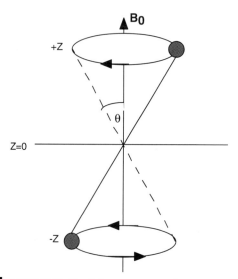

FIGURE 2.10 Bulk magnetization illustration.

back to equilibrium and is given by T_1 (longitudinal relaxation time). "Transverse" relaxation causes the transverse magnetization to be reduced to zero and is given by T_2 (transverse relaxation time). In most materials $T_2 < T_1$, and T_1 is usually shorter for liquids than for solids. On the other hand, T_2 is generally greater for liquids than for solids.

2.4. MAGNETIC RESONANCE IMAGING HARDWARE DESIGN

The basic components of MRI hardware are shown in Figure 2.11. The magnet in MRI is responsible for generating the static magnetic field B_0. The gradient

MAGNETIC RESONANCE IMAGING HARDWARE DESIGN

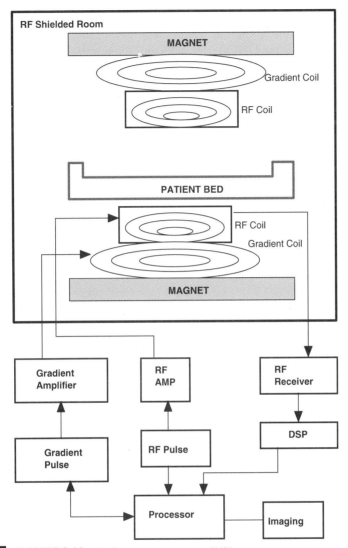

FIGURE 2.11 Hardware representation of MRI.

coils create a gradient which is superimposed upon the main field B_0. A strong magnetic field provides a better SNR and better resolution in both frequency spatial domains. Most clinical imaging systems have field strengths around 2 T. Some functional MRI systems have 3 T or 4 T main magnets.

The primary requirements of the main magnet is that its field be uniform. In most cases, the magnetic field is not uniform and for this reason "shim" coils are frequently employed. The shim coils are a set of coils built to produce a field that is polarized in the same direction as the main field of known spatial dependence. Therefore, if the main magnet's nonuniformity is known, the shim coils can be set to carry gradients that cancel (using superposition) the inhomogeneous components of the main field.

Most magnetic fields are generated using permanent magnets. These magnets are simple and affordable. Their fringing fields are also small. Permanent magnets can also come in different sizes and shapes. In MRI applications with permanent magnets, the patient is placed between the two poles of the magnets. However, temperature drift of permanent magnets is an issue since the RF frequency is usually not controlled by any type of feedback and would therefore not remain at the Larmor frequency if the temperature changes after calibration.

For magnetic fields greater than 0.5 T, superconducting magnets must be used. The coil windings are made from an alloy of niobium–titanium and are cooled to a temperature below 12 K using liquid helium (boiling point of 4.2 K). These fields are very strong and homogeneous.

The Larmor frequency depends on the strength of the magnetic field B_0. Strong gradients for imaging technique may require the help of gradient pulsing. The duration of these pulses must be of the order of T_2 to be effective. Because gradient coils have a natural self-inductance, they cannot be switched on or off simultaneously. Coils that have large inductances can be driven to steady state quickly at the expense of power by using other driver circuits to compensate for the power needed. Since the gradient coils can induce eddy currents in the rest of the magnet structure, to compensate, shielding must be used. Furthermore, the desired gradients, which are typically linear, must provide detailed spatial information about the sample. Gradients for most modern imaging schemes can be produced in any of the three spatial directions without physically rotating the gradient coils.

The sample under test must be irradiated with an RF field also known as the B_1 field. This is done to tip the magnetization away from the equilibrium position and generate a detectable NMR signal. The RF fields are produced by a transmitter and an RF coil. The transmitter determines pulse shape, duration, power, and timing. The RF coil is responsible for coupling the energy given by the transmitter to the nuclei. To generate an RF pulse, the transmitter uses first a frequency synthesizer to achieve the defined frequency. A waveform generator creates a user-defined pulse shape, which is then mixed with the pure tone and this creates an RF pulse. The pulse or a sequence of several pulses is repeated at a repetition rate. There are basically two types of pulses used in NMR machines: the hand pulses, which are typically broad, and rectangular pulses. The other type of pulse is known as soft pulses, which are often sinc shaped to provide frequency selectivity. In soft pulsing, it is important that the center frequency is close to the Larmor frequency.

To receive a good signal, RF coils are required to couple the energy from nuclei in precedence to a receiver. The RF coil must be responsive to frequencies within the general band of the Lamor frequency of interest. In human imaging, the wavelength is generally of the same order of magnitude as the sample. Often, a coil design known as the "birdcage" is used. This coil combines lumped capacitors with distributed inductance to form a volume resonator. If the resonance of the coil is designed well, it should increase the efficiency of the coil at the operating frequency. The lumped capacitors provide a method for the resonator to store energy, as shown in Figure 2.12.

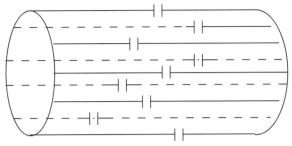

FIGURE 2.12 Lumped capacitor model for the coil in the MRI resonator.

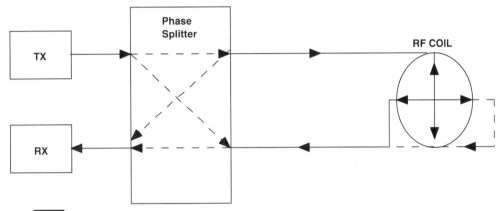

FIGURE 2.13 Pulsing RF coils with perpendicular fields.

The coils must generate a very uniform B_1 magnetic field. Any inhomogeneity will introduce a distortion in the images. The requirement of a uniform field places a burden on the coil design. Therefore, it is desirable to use a coil that allows quadrature excitation and detection (two RF coils are needed, one for transmitting and one for receiving). Quadrature excitation is capable of generating and receiving circularly polarized fields. To accomplish this, the transmittal power and received power must be split equally into two channels, with one of the channels having a 90° phase delay. The two channels are then fed to the two inputs of the RF coil to produce fields that are perpendicular to each other as shown in Figure 2.13.

2.5. PULSING AND NMR IMAGING

One "pulsing" scheme commonly used today is called "spin echo" sequences. This sequence provides a 90° pulse followed by a time delay $T_d/2$ (T_d is time delay), a 180° pulse, and another time delay, as shown in Figure 2.14. The term T_r denotes the length of a frame.

After the 90° pulse, the magnetization lies in the transverse plane if the pulse of 90° is applied along the x axis. It results in a magnetization aligned along the y axis and coherence has been established. The nuclei under the influence of stronger fields rotate faster and hence have some net rotation at the mean

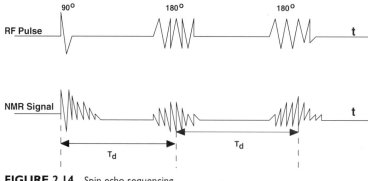

FIGURE 2.14 Spin echo sequencing.

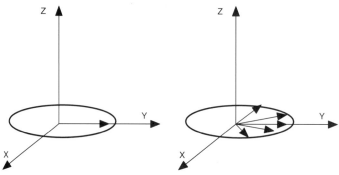

FIGURE 2.15 Nuclei under the influence of a magnetic field.

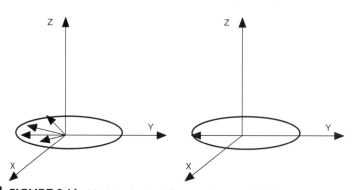

FIGURE 2.16 Nuclei under the influence of a magnetic field.

Larmor frequency. The nuclei under the influence of a weaker magnetic field will rotate slowly, one lag behind, and experience a negative sensed net rotation in the frame, as shown in Figure 2.15.

If we now apply a 180° pulse, also on the x axis, each individual magnetic moment will be rotated 180° about the x axis (Figure 2.16). Now the faster nuclei are behind the mean nuclei and the slower nuclei are ahead. This means that the faster nuclei eventually will cross the path of the slower nuclei and coherence is again established (Figure 2.16).

PULSING AND NMR IMAGING

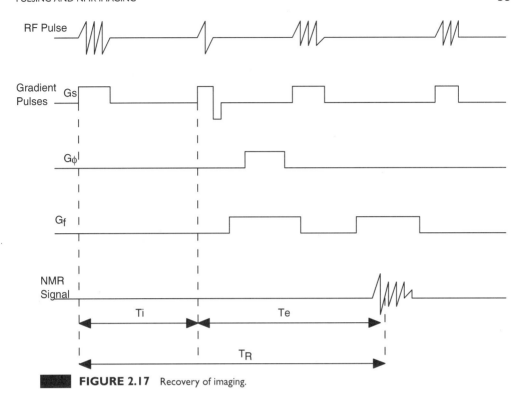

FIGURE 2.17 Recovery of imaging.

The peak value of the "echo" signal is given by

$$E_0 = (\text{free induction decay}) \exp\left(-\frac{T_d}{T_2}\right) \quad (2.24)$$

from which the value of T_2 can be obtained.

The inversion recovery imaging that uses the spin echo is shown in Figure 2.17, where G_s is the slice selection gradient, G_f is the required readout gradient, and G_ϕ is the phase-encoding gradient. The sequence starts with an RF pulse at 180° in conjunction with a slice selection gradient. The remainder of the sequence after a time T_i is the same as the spin echo imaging sequence. The spin echo sequences employ 90° and 180° RF pulses.

The inversion recovery imaging technique can be performed in conjunction with the two-dimensional (2D) Fourier transform technique. We first define the slice to be imaged out of a three-dimensional (3D) object by applying a steep gradient perpendicular to this slice. Next it is necessary to phase encode each of the longitudinal directions of the 2D slice. This will be accomplished by applying a gradient pulse along each of those directions. Usually, for the first acquisition, the pulse is halted once nuclei on the edges of the sample have accumulated a pulse difference of 180°. When the cycles are repeated larger phases shifts are deliberately induced in each cycle in a progressive manner. This is accomplished to increase the resolution in the longitudinal direction, which for this method depends on the number of the acquisitions.

REFERENCES

Brown, Mark A., and Richard C. Semelka. 1999. "Magnetic resonance abbreviations, definitions, and description—a review." *Radiology*.

Chien, Daisy, and Robert R. Edelmen. 1991. "Ultrafast imaging using gradient echoes, magnetic resonance." *Quarterly* 7, 31–56.

De Yoe, Edgar A., Pater Bandettini, Jay Neitz, David Miller, and Paula Winans. 1994. *J. Neurosci. Meth.* 54, 171–187.

Jin, Jianming. 1999. *Electromagnetic Analysis and Design in Magnetic Resonance Imaging*, CRC.

Lauterbur, P. C. 1973. "Image formation by induced local interactions: examples employing nuclear magnetic resonances." *Nature* 242.

Mansfield, P., and P. G. Morris. 1982. *NMR Imaging in Biomedicine*. Academic Press, New York.

Slitcher, C. P. 1986. *Principles of Magnetic Resonance*, 3rd ed. Springer Verlag, New York.

3
PARTICLE ACCELERATOR DESIGN

3.1. INTRODUCTION

The earliest radiation sources known to humans were provided by gas-filled x-ray tubes. In and around 1913 W. D. Coolidge developed a vacuum tube with a hot tungsten cathode as the preliminary work for present-day x-ray techniques. The earliest tubes of this type, manufactured for general use, operated at a peak voltage of 140 kV with a 5-mA current. By 1937 x-ray tubes used in radiation therapy operated at 400 kV with 5 mA of current, but the x rays generated by these tubes were fairly soft with very little penetration beyond the skin.

In the early half of the 1950s the era of cobalt sources for radiotherapy began with cobalt-60. The energies emitted were in the range of 1.17 and 1.33 MeV and this allows for better penetration than x rays (Figure 3.1).

The first of the accelerators for medical therapy was a high-voltage accelerator developed in 1922 by R. Van de Graff. The first hospital-based accelerator was a 1-MeV air-insulated machine installed in Boston in 1937. The second machine, a pressure-insulated 1.25-MeV version, was installed in 1940 at the Massachusetts General Hospital. Eventually 40 such accelerators were installed, but they were discontinued in 1959 because of their ever-increasing sizes and their high cost.

A dramatic increase in the photon radiation energy was made possible in 1943 with the development of a new accelerator known as the betatron. The first patient was irradiated in 1949 with x rays generated by a 20-MeV electron gun from a betatron in Urbana, Illinois. By 1977, 45 betatrons were

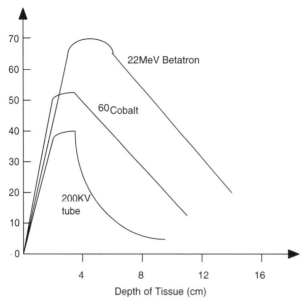

FIGURE 3.1 Penetration of different x rays for therapy.

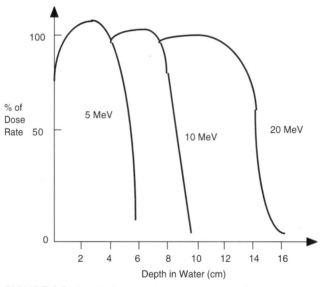

FIGURE 3.2 Depth–dose distribution per tissue depths.

in operation in the United States alone. Betatrons deliver x rays with such properties as skin sparing, greater depth dose, and less side scatter than those of the radiation generated by x-ray tubes and radioisotope sources. In Figure 3.2 we observe a depth–dose distribution that shows that at a given tissue depth, the dose absorbed using a betatron is much higher than a 200-kV x-ray and ^{60}Co source. The main disadvantage of the betatron is the relatively low intensity of the x-ray beam as well as the small treatment field area.

TABLE 3.1 Some parameters of photon therapy devices

Beam	Accelerator	Advantage	Disadvantage
Superficial x rays (10 to 100 kV)	Transformer + vacuum tube	Low depth dose for superficial lesions	High side scatter (sloppy beam)
Deep x rays (180 to 450 kV)	Transformer + vacuum tube	Improved depth dose for deeper or thicker lesions	High skin dose; high side scatter; depth dose inadequate for deep lesions of the trunk
1- to 2-MV x rays	Van de Graaff	Skin sparing; improved depth dose; reduced side scatter; little penumbra	Relatively fixed machine; awkward in other than vertical or horizontal treatment
Cobalt-60 teletherapy	Radioactive source, no accelerator	Skin sparing; isocentric mount; depth dose similar to 1- to 2-MeV x rays	Large penumbra; decay of source
4- to 16-MV x rays	Electron linear accelerator	Skin sparing; depth dose; isocentric mount; small penumbra; little side scatter; high dose rate	Depth dose less than with higher energies
20- to 35-MV x rays	Electron linear accelerator (linac)	Similar to 4- to 16-MV; greater depth dose	High cost; neutron production in patient and treatment room
	Betatron	Similar to linac	Similar to linac; relatively fixed unit; dose rate may be low

The era of RF linear accelerators began in the mid-1960s and it eventually became the dominant force in the world market for medical accelerators. The limitation of betatrons and the capability of RF linear accelerators to provide a high dose rate have contributed to the rise of linear accelerators and their dominant role among the world's radiotherapy machines. The trend to use even higher photon energies was particularly noticeable in the early 1970s when production of 45- to 50-MeV accelerators began. This early enthusiasm was soon revised by unfavorable side effects encountered in this high-energy range, especially those caused by harmful neutrons generated in the patient's body. This eventually led to the use of lower energies not exceeding the 20- to 25-MeV range. Table 3.1 lists the advantages and disadvantages of generators used in photon therapy devices.

Electron therapy is a radiotherapy that involves the direct use of accelerated electrons. The beams are of low energy (5–10 MeV) and have a flat peak and a fast drop on the depth–dose curve.

Like photon therapy, electron treatment also employs electrons with various energies, depending on the treatment. Table 3.2 lists routine applications of electron beams.

TABLE 3.2   Routine clinical applications of the electron beam

Type of cancer	Tissue/organs affected	Application dose
All skin cancers	Lip, cheek, temple; lesions critically located; eyelids, nose, ear; lymphomas of the dermis, sarcomas on the dorsum	6–8 MeV; if the lesion is thicker than 2 cm, 10 MeV
Breast and chest wall	Lesions on chest wall and lymphatics; after removal of the primary lesion	8–14 MeV
Upper respiratory and digestive tract	Oral cavity: floor of the mouth; undersurface of the tongue; often electrons + interstitial Intraoral cone: electrons alone; large lesions of gingival: boost with electrons after photons (external)	8 MeV
Neck	Salivary glands: electrons + photons; lymphatic of the neck: multiple fields of various energies	10–18 MeV
Bladder, lungs, cervix, colon and rectum		Energies above 18 MeV; problems with inhomogeneities

3.2. HEAVY PARTICLES

Radiation with heavy particles involves the use of protons, heavy ions, and alpha particles as well as pi-mesons and neutrons. Heavy particles can provide much better results when radiating the target area (i.e., corresponding directly to the tumor tissue), while the energy absorbed by normal tissue around the tumor is minimized. Neutron and proton therapies are becoming more popular but the costs are still much higher when compared with photon and electron therapy.

In radiobiological studies, charged particle accelerators have advantages over other sources of ionizing radiation for medical treatment of cancer: (1) considerably higher energies can be achieved, (2) easily controlled accelerated particle energies constitute an additional parameter in some experiments, (3) the variety of mass numbers of the bombarding ions provides a variety of treatment options, and (4) beams with higher currents eliminate the risks of chemical contamination.

3.3. PARTICLE ACCELERATORS

Electrical acceleration of charged particles is the basis of RF linear accelerators. These particles can range from electrons to ions. As the mass of the particle increases, the ratio of the charge (q) to the mass (m), that is, q/m, decreases, and the efficiency of the acceleration process deteriorates. The fundamental properties of the charged particles most commonly used in biomedical accelerators

TABLE 3.3 Fundamental particles in commonly used particle linear accelerator

Particles	Rest mass (Kg)	Rest energy (MeV)	Charge ×10⁻¹⁹ C
Electron	9.11×10^{-31}	0.511	−1.602
Proton	1.67×10^{-27}	938.3	1.602
Deuteron	3.342×10^{-27}	1877	1.602
α Particles	6.664×10^{-27}	3733	3.204

are given in Table 3.3. Electrons with a high q/m ratio are used in electron or photon therapy, and they are widely used for sterilization purposes. Protons are the nuclei of the fundamental hydrogen atom. For biomedical purposes, they are used directly in biomedical radiation treatment, and after conversion into mesons they can be used in neutron therapy. In addition, they are the particles most frequently employed in the production of medical radioisotopes. Deuterons are the nuclei of deuterium, a hydrogen isotope with a mass of 2. Deuterons are employed in acceleration because they have a large cross section for several nuclear reactions, such as those involved in the production of radio nuclides. Alpha particles are the nuclei of the helium atom ^{4}He.

When a normally electrically neutral atom is subjected to a strong atomic interaction (i.e., a collision with another atom) it acquires a positive charge numerically equal to that of the remover electrons and it becomes a positive ion. The mass of the ion is given by $m_i = (Z + N)M_n = AM_n$, where Z is the atomic number, $N = A - Z$ is the number of neutrons in the nucleus, A is the mass number, and M_n is the mass of the nucleon. The kinetic energy of the ion particle is given by

$$E_i = \frac{M_0 c^2}{\sqrt{1 - \beta^2}} - M_0 c^2 \tag{3.1}$$

where $\beta = v/c$ is the relative velocity, v is the particle velocity, M_0 is the rest mass of the particle, and c is the speed of light. When the velocity of the particle becomes comparable to that of the speed of light, the particle is said to be behaving as a relativistic particle with a mass given by

$$m = \frac{M_0}{\sqrt{1 - \beta^2}} \tag{3.2}$$

Applications of various charged particles in biomedical applications and their corresponding typical energy ranges are given in Table 3.4.

From Eq. (3.1) a charged particle acceleration is given by the expression

$$E_i = \frac{M_0 c^2}{\sqrt{1 - \beta^2}} - M_0 c^2 = \vec{E}(z)\, dz \tag{3.3}$$

where $\vec{E}(z)$ is the electric field, and z is the length of the acceleration path; the simplest way to accelerate a charged particle is to use two electrodes. A charged particle in the space between the two electrodes is repelled by the electrode of

TABLE 3.4 Biomedical applications and typical energy ranges for various charged particles

Particle	Energy range	Application
Electrons	6 to 25 MeV	Routine electron and photon radiotherapy
	2 to 10 MeV	Radiosterilization, pasteurization, food preservation
	Several hundred MeV to several GeV	Production of synchrotron radiation for angiography and analysis
Protons	3 to 10 MeV	Analytical methods
	Several tens of MeV	Neutron production in experimental therapy
	Several tens of MeV to 250 MeV	Proton radiotherapy
	10 to 100 MeV	Meson production in experimental therapy Medical radioisotope production
Deuterons	7 to 20 MeV	Medical radioisotope production
	5 to 20 MeV	Analytical methods
Heavy ions	70 to 700 MeV/nucleon	Experimental radiotherapy

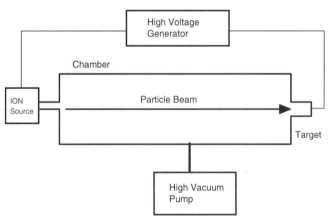

FIGURE 3.3 Basic block diagram of a linear accelerator.

the same sign and is attracted by the electrode of the opposite sign. An example of such a device is shown in Figure 3.3 as the basic schematic of a high-voltage linear accelerator.

In a voltage linear accelerator, the kinetic energy of the particle is given by

$$E_i = qV \qquad (3.4)$$

for an electron $q = e$ and $E = $ eV where $1\,\text{eV} = 1.6 \times 10^{-19}$ J on the energy gained by $1e$ when a potential difference of 1 V is used.

3.4. LINEAR RF ACCELERATORS

The great advances in radioengineering, radar, and microwave technology have established a solid background for charged particle methods based on the electric field $\vec{E}(z)$. These acceleration methods are suitable for both light particles, such as electrons, and heavy ions since it is possible to achieve energies between mega- and giga-electron-volts.

The first of such accelerators was known as the Alvarez structure. It consisted of a set of resonators that have an RF voltage of the same phase applied to them (see Figure 3.4a). Inside each resonator exists a potential distribution, as shown in Figure 3.4b. Therefore, the acceleration takes places in the resonator gaps. The edges of the adjacent resonators are equipped with cylindrical drift electrodes, also of variable length, matched in such a way as to screen the particles against the decelerating effect of the field.

The basic schematic of an RF accelerator for electrons is shown in Figure 3.5. The electron beam is produced in the accelerator gun with a heated cathode operating in a triode circuit. The anode is powered by several hundreds kilovolts to accelerate the electrons to a velocity of about $0.5c$. The grid

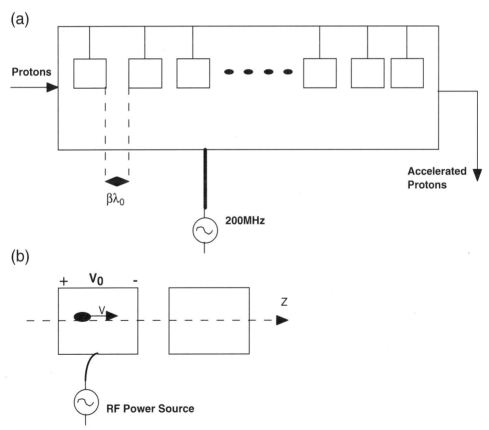

FIGURE 3.4 Example of the Alvarez structure accelerator. (a) Overall structure, (b) accelerator chamber.

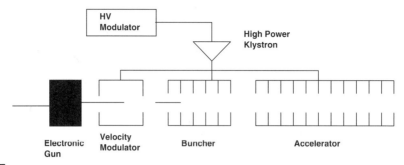

FIGURE 3.5 Schematics of an RF linear accelerator.

to control the electron beam is usually negatively polarized with respect to the cathode so that the electrons cannot go through it. The pulses that activate the grid range between 1 and 10 μs at a rate between 500 and 1000 pps. The single pulse current amplitude can be as high as 10 A.

The RF supply is produced in generators equipped with either magnetrons or klystrons operating in the S-band (1.55–5.2 GHz). The beam of electrons produce by the $\bar{e}$ gun arrives first at a prebuncher, which is a single, low-Q resonator, supplied with about 2 kW of RF power at the fundamental frequency. The prebuncher modulates the velocity of the electrons. Some RF linear accelerators provide an additional buncher system with a transverse electric field operating as an electric gate, letting through single $\bar{e}$ bunches every few nanoseconds. The electron bunches at its output attain an energy of about 250 keV, which correspond to a velocity of $0.75c$. The beam formed is then directed to successive accelerating sections. Such sections are very large (thousands of meters).

3.5. PARTICLES ACCELERATED BY A MAGNETIC FIELD

If a particle with a mass $m = Am_n$ and charge q travels with a velocity v in a direction perpendicular to the lines of the magnetic field of induction $\vec{B}$, it is subjected to the Lorentz force F perpendicular to the direction v and $\vec{B}$ (see Figure 3.6):

$$F = vBq \tag{3.5}$$

Since the centrifugal force $F_c = mv^2/r$, we can equate these two quantities $mv^2r = vBq$ and obtain

$$Bv = \frac{M_n v A}{q} \tag{3.6}$$

The product $B \times r$ is called the particle magnetic rigidity. It follows from this last equation that the higher the ratio of the charge to the mass number, the lower is the value of B needed for an orbit with the same radius r. The particle frequency of revolution is given by

$$f = \frac{v}{2\pi r} = \frac{qB}{2\pi AM_n} \tag{3.7}$$

PARTICLES ACCELERATED BY A MAGNETIC FIELD 65

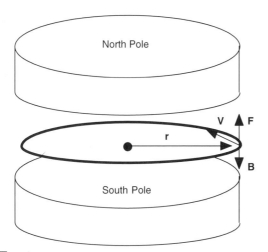

FIGURE 3.6 Particle under the effects of magnetic forces.

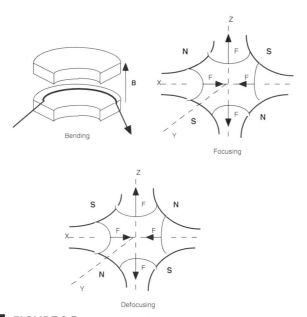

FIGURE 3.7 Focusing and defocusing of beams in magnets.

At room temperature, there are limitations on the materials for magnets and the product Bv can be increased only up to a certain point without making the magnets too large and heavy. Therefore, superconducting magnets whose excitation windings operate at the temperature of liquid helium have been used with a great reduction in weight, size costs, and power supply. Focusing and bending the beams can be accomplished with different types of magnet constructions, as shown in Figure 3.7.

3.6. SYNCHROTRONS

Figure 3.8 shows a schematic diagram of a proton synchrotron. Preaccelerated particles are injected into the accelerating chamber, which is equipped with many electromagnets for bending the protons into a circular path. In synchrotrons, the acceleration is implemented with resonators supplied with an RF voltage in a way very similar to RF linear accelerators. In an electron synchrotron, the orbital frequency of electrons is approximately constant during the total acceleration cycle, due to the fact electrons of relatively low energies achieve velocities near the speed of light. Therefore, the accelerating voltage frequency can also be kept constant.

In proton synchrotrons, the velocity of protons on heavy ions that slowly gain speed varies widely during the acceleration cycle, and this requires frequency modulation in addition to the magnetic field alternating with time.

In a synchrotron, the particles are accelerated in an orbit with constant radius. As the particle energy arises due to the acceleration, the magnetic field energy that also arises in the magnetic field that keeps them in a stable orbit must also increase with time. The magnetic field then remains stable for a while (about 1 s), known as the waiting period, during which the beam of accelerated protons is gradually extracted.

In the final stage of the cycle (about 0.5 s) the magnetic field falls from the maximum to its output value. In large proton synchrotrons, the operation ranges from several seconds to several tens of seconds. Therefore, the repetition frequency of the subsequent acceleration cycles is rather small. The synchrotron method is at present the only one that can afford proton acceleration energies of more than 100 GeV and can hold great promise for biomedical treatment.

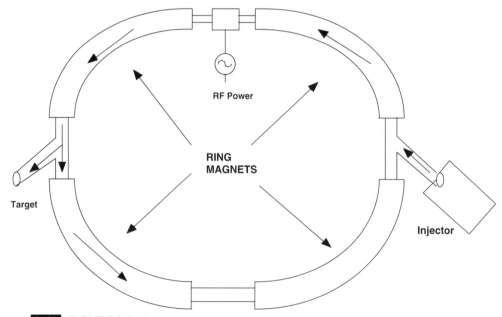

FIGURE 3.8 Synchrotron diagram.

For proton therapy, the maximum energy required is about 250 MeV. Heavy ion therapy can also be conducted with synchrotrons.

3.7. ACCELERATOR HARDWARE

Accelerators used in radiotherapy are powered by an RF energy from microwave tubes, such as magnetrons and klystrons, and operate in the *s*-band with a frequency of 3 GHz. The operating principle of the RF klystron is shown in Figure 3.9. All the electrons passing the first cavity gap at zero of the gap voltage (or signal voltage) pass through with unchanged velocity; those passing through the positive half-cycles of the gap voltage undergo a decrease in velocity. As a result, the electrons gradually bunch together as they travel down the drift space. The variation in electron velocity in the drift space is known as velocity modulation.

The density of the electrons in the second cavity gap varies cyclically with time. The electron beam contains an ac component and is said to be current modulated. The maximum bunching occurs around midway between the second cavity grids during the retarding phase. Thus the kinetic energy is from the electrons to the field of the second cavity. The electron energy from the second cavity is reduced in velocity and then terminates at the collector.

When electrons are first accelerated by the high dc voltage V_0 before entering the buncher grids, their velocity is uniform:

$$v_0 = 0.593 \times 10^6 \sqrt{V_0}\, \text{m/s} \tag{3.8}$$

When a microwave signal is applied to the input terminal, the gap voltage between the buncher grids appears as

$$V_s = V_1 \sin \omega t \tag{3.9}$$

where V_1 is the amplitude of the signal and $V_1 \ll V_0$ is assumed.

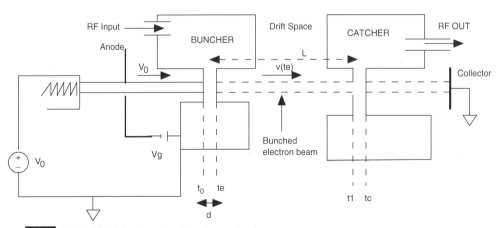

FIGURE 3.9 Example of accelerator hardware.

The modulated velocity in the buncher cavity in terms of the time to enter is given by

$$v(t_1) = \sqrt{\frac{ze}{m} V_0} \left[1 + \frac{\beta_i V_i}{V_0} \sin\left(wt_0 + \frac{\theta_g}{2}\right) \right] \quad (3.10)$$

where θ_g is the average gap transit angle $= (\omega d)/V_0$, d is the buncher gap distance (see Figure 3.9), m is the mass of an electron, and $\beta_i = [\sin(\theta_g/2)]/\theta_g$ is the beam coupling coefficient of the input cavity gap.

If we approximate further by $\beta_i V_1 \ll V_0$, then

$$v(t_1) = V_0 \left[1 + \frac{\beta_i V_1}{2 V_0} \sin\left(wt_1 - \frac{\theta_g}{2}\right) \right] \quad (3.11)$$

Once the electrons leave the buncher cavity, they drift with a velocity $v(tr)$ along in the field-fire space between the two cavities.

The optimum distance L (see Figure 3.9) at which the maximum fundamental component of current occurs is given by

$$L_{\text{opt}} = \frac{3.682\, v_0 V_0}{w \beta_i V_1} \quad (3.12)$$

and the beam current of the catcher cavity is given by

$$I_c = I_0 + \sum_{n=1}^{\alpha} 2\, I_0 J_n(nX) \cos n\omega \quad (t_1 - \tau - T_0)$$

where

$$X = \frac{\beta_i V_1}{2 V_0}(2\pi N)$$

N = the number of electron transit cycles in the drift space
I_0 = the dc current
$T_0 = L/V_0$ (dc transit time)
$\tau = d/V_0$
$Jn(nX)$ = nth-order Bessel function of the first kind

The output power delivered to the catcher is given by

$$P_{\text{out}} = \frac{\beta_0 I_2 V_2}{2} \quad (3.13)$$

$$I_2 = 2\, I_0 J_1(x)$$

where V_2 is the fundamental component of the catcher gap voltage, β_0 is the beam coupling coefficient of the catcher gap, and $\beta_i = \beta_0$ if the cavities are identical.

MAGNETRONS

3.8. MAGNETRONS

All magnetrons consist of some form of anode and cathode operated in a dc magnetic field normal to a dc electric field between the anode and the cathode. Because of the cross-fields between the anode and the cathode, electrons emitted from the cathode will move in curved paths. If the dc magnetic field is strong enough, the electrons will not arrive at the anode but return instead to the cathode. An illustration of a circular magnetron is shown in Figure 3.10. In circular magnetrons, the electrons follow cycloidal paths in the cathode anode space.

Whether the electron will just graze the anode and return to the cathode depends on the relative magnitude of V_0 and B_0. This is a condition called the

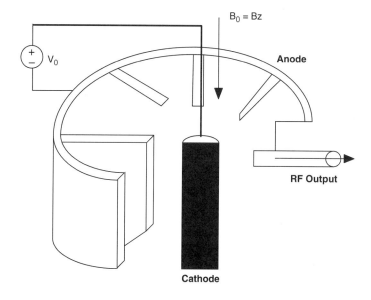

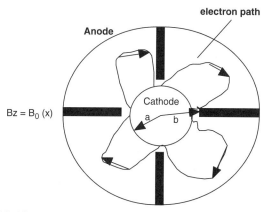

FIGURE 3.10 Circular magnetron illustration.

cutoff condition. The cutoff magnetic flux is given by

$$B_{\text{cutoff}} = \frac{(8V_0 m/e)^{1/2}}{b(1 - a^2/b^2)} \tag{3.14}$$

This means that if $B_0 < B_{\text{cutoff}}$ for a given B_0 the electrons will not reach the anode. The cutoff voltage is given by

$$V_{\text{cutoff}} = \frac{e}{8m} B_0^2 b^2 \left(1 - \frac{a^2}{b^2}\right)^2 \tag{3.15}$$

This means that if $V_0 < V_{\text{cutoff}}$ for a given B_0, the electrons will not reach the anode. The angular frequency of the circular motion of the electron is given by

$$\omega = \frac{eB_0}{m} \tag{3.16}$$

3.9. ACCELERATOR ARCHITECTURE

A general system diagram of an accelerator is shown in Figure 3.11. The operation begins with a modulator that generates high-voltage pulses delivered to the RF generator, which could be either a magnetron or a klystron. The modulator also supplies pulses to the electron gun. The length of the pulses is of the order of a few microseconds. In some cases, the RF generator is coupled to the accelerating structure with the isolator. The isolator is used to prevent energy reflected from the structure from reaching the RF generator, which could cause a loss of phase stability and an electrical breakdown.

The accelerating structure is also coupled to the RF generator by means of an automatic frequency control (AFC) servo. The detection for the AFC is supplied by two additional internal resonant cavities tuned above or below the resonant frequency. When the instantaneous operating frequency changes in either direction, the AFC system causes the magnetron to be tuned to the postulated operating frequency.

The treatment head shown in Figure 3.11 is one used for x-ray therapy. The electron beam strikes a target and becomes converted to x rays. The x rays pass through a primary collimator and then enter a conical flattering filter. Ion chambers and variable collimators are located adjacent to the primary collimator. The treatment head can also contain elements that modify the beam shape and an optical system that determines the field size and the skin–target distance, if the accelerator is intended for both x rays.

In Figure 3.12, we observe that the control system electron is under operator control.

In accelerators for routine radiation therapy, the design currently used is that shown in Figure 3.13. If the magnetron is employed as a source of RF energy, all electronic components as well as the acceleration structure and the treatment head are contained in a gantry that can rotate 360° around the rotation axis. In Figure 3.13, the patient lies on the treatment couch, which has a great range of linear and rotational movements. The tumor center must be

ACCELERATOR ARCHITECTURE

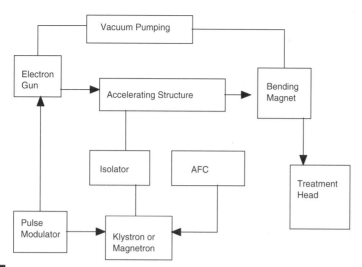

FIGURE 3.11 Block diagram of an RF linear accelerator (linac).

located at the isocenter, a point defined by the orthogonal intersection of the axis of rotation of the head and the therapy beam axis.

3.9.1. Traveling Wave Accelerator System

This system is shown in Figure 3.14. The magnetron and the electron gun are modulated by periodic pulses from a modulator. The RF energy generated in the magnetron passes from the left-hand side of the accelerating structure through an isolator and the RF waveguide window. The isolator allows the RF energy to pass from the magnetron to the structure, but it prevents the possible return of the energy reflected from the magnetron. The RF energy passes through the structure, where it is mostly transferred to the beam of accelerated electrons. The residual energy is transferred to the load resistance coupled to the right-hand end of the accelerating structure through an RF window.

3.9.2. Standing Wave Accelerator

A system of this kind has a modulator, a klystron generator, and other auxiliary circuits, each housed in a different stationary cabinet. The accelerating structure is responsible for accelerating electrons that are delivered with energy of up to 10 MeV in the x-ray modality. These electrons are delivered to the magnet system, which provides a chromatic beam bending 270°, and then are converted to x rays in a target located in the gantry. The structure can generate electrons with energies of 6, 9, 12, 15, and 18 MeV. The structure is powered by RF energy via a rotating RF joint located between the movable frame and the stationary cabinet.

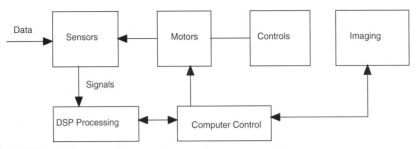

FIGURE 3.12 Control system electronics for an x-ray machine.

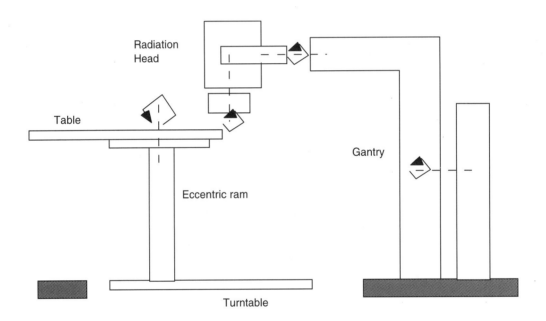

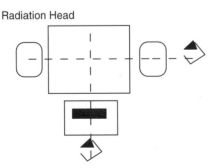

FIGURE 3.13 The rotary gantry radiation table for routine therapy.

ACCELERATOR ARCHITECTURE

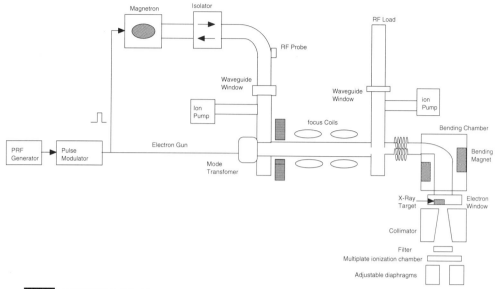

FIGURE 3.14 Travelling wave radiation therapy accelerator.

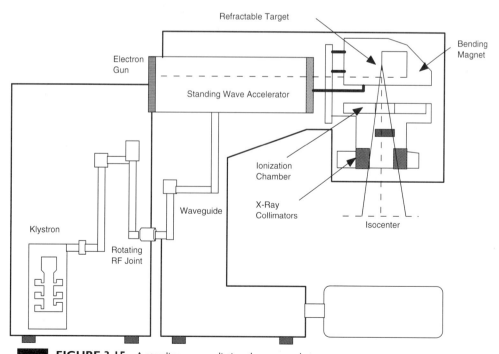

FIGURE 3.15 A standing wave radiation therapy accelerator.

REFERENCES

Bewley, D. K. 1985. "The 8MeV linear accelerator at the MRC Cyclotron Unit." *Br. J. Radiol.* 58.
Craddock, M. K. 1987. "Cyclotrons in 1986." *Proc. 11th Conf. on Cyclotrons and Their Appl.*, Ionics, Tokyo.
Hamm, R. W. 1987. "Linacs for medical and industrial applications." *Proc. 1986 Linear Accelerator Conference.*
Hendee, William, R. 1995. "X-rays in medicine." *Physics Today* 51–56.
Heusinkveld, R. S., *et al.* 1975. "Accelerated particles in radiation oncology." *IEEE Trans. Nucl. Sci.* NS-22, 1211–1215.
Jianming Jin. 1999. *Electromagnetic Analysis and Design in Magnetic Resonance Imaging.* CRC Press.
Lam, G. K. Y., *et al.* 1985. "Cancer radiotherapy using negative pi-mesons." *Nucl. Instrum. Meth.* B10/11, 1096–1101.
"LANL Report." 1991. *Proc. of the 1988 Linear Accelerator Conf.* Los Alamos.
Lauterbur, J. P. "The basics of MRI." www.cis.rit.edu/htbooks/mri
Loew, G. 1984. *Electron Linear Accelerators and Stretcher.* SLAC Publ-3343, Standford, May.
Loew, G., and R. Talman. 1983. *Elementary Principles of Linear Accelerators.* SLAC Publ-3221, Standford.
Magin, L. R., *et al.* 1997. "Miniature magnetic resonance machines." *IEEE Spectrum*, October.
Mansfield, P. and P. G. Morris. 1982. *NMR Imaging in Biomedicine.* New York, Academic Press.
Nunan, C. S. 1985. "Design and performance criteria for medical electron accelerators." *Nucl. Instrum. Meth.* B10/11, 881–887.
Raju, M. R. 1980. *Heavy Particles Radiotherapy.* Academic Press, New York, London.
Slater, J. M., *et al.* 1988. "Development of a hospital based proton beam treatment center." *Int. J. Radiat. Oncol. Biol. Phys.* 14, 761–775.
Slichter, C. P. *Principles of Magnetic Resonance.* (3rd ed.) New York, Springer Verlag.
Wehrili, Felix, W. 1992. "The origins and future of nuclear magnetic resonance imaging." *Physics Today*, June.
Young, S. W. 1984. *Nuclear Magnetic Resonance Imaging: Basic Principles.* New York, Raven Press.

4

SENSOR CHARACTERISTICS

4.1. SENSOR PARAMETERS

A sensor produces an electrical signal from its input to its output. In principle, a sensor can be regarded as a black box where the important thing is the relationship between the input signal and the output signal. The theoretical relationship between input and output signals is established through a transfer function. This function establishes the dependence between the output electrical signal S and the stimulus input I. These transfer functions may be linear, as

$$S = a_0 + a_1 I^n \qquad (4.1)$$

where n is a constant number, or they can be nonlinear, as

$$S = a e^{nI}$$
$$S = a + a_1 \ln I$$

Other characteristics of sensors are:

Accuracy: In reality, this means inaccuracy. Accuracy is measured as the ratio of the highest deviation of a value represented by the sensor to the ideal value. The deviation can be described as a difference between the ideal input value and a value that was converted by the sensor into voltage and then, without error, was converted back.

*Thanks are expressed to Analog Devices for the usage of some of their components publications.

Calibration Error: This is the inaccuracy permitted by a manufacturer when a sensor is calibrated in the factory. This error is of a systemic nature, which means that it is added to all possible real transfer functions. This shifts the accuracy of transduction for each stimulus by a constant. The error is not necessarily uniform over the range and may change, depending on the type of calibration error.

Hysteresis: This is a deviation of the sensor's output at a specified point of input signal when it is approached from opposite directions as shown in Figure 4.1, a phenomenon in which the state of a system does not reversibly follow changes in an external parameter. For example, in a displacement sensor, it's the difference in output reading (V_{out}) obtained at a given point when approaching that point from upscale and downscale readings (input).

Nonlinearity: Nonlinearity error is specified for sensors whose transfer function can be approximated by a straight line. A nonlinearity is a maximum deviation (D) of a real transform function from the straight line. Nonlinearity is usually specified as a percentage of span in terms of the straight line or measured value. One way to specify nonlinearity is through the method of least squares, as shown in Figure 4.2. In this

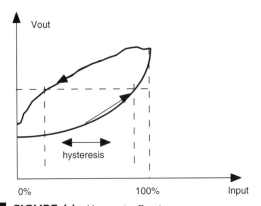

FIGURE 4.1   Hysteresis effect in sensors.

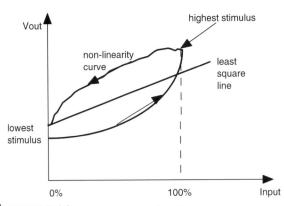

FIGURE 4.2   Least-squares method for hysteresis.

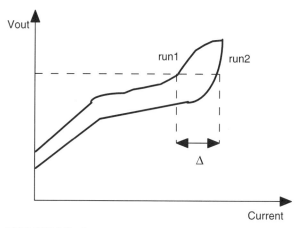
FIGURE 4.3 Error correction in reliability.

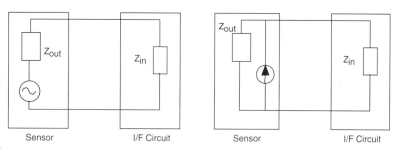
FIGURE 4.4 Types of sensor connections to I–F circuits.

figure the least square line is the "Real" transform function. The nonlinearity curves (above and below the line) represent the percentage of deviation from that linearity. The lowest and highest stimulus points in Figure 4.2 represent how much is the nonlinearity at the points where the inputs are of lowest and highest magnitude, respectively.

Saturation: Every sensor eventually has an operating limit. Even when the sensor behavior is linear, there will be some high- or low-input level at which the sensor output will not respond.

Repeatability: This error is caused by the inability of a sensor to represent the same value under identical conditions, as shown in Figure 4.3.

Dead Band: This is the insensitivity of a sensor in a specific range of input signals. The output will remain the same over a certain dead band zone.

Resolution: This describes the smallest increments of stimulus that can be sensed. When the input continuously varies over the range, the output signals of some sensors will not be very smooth and the output may change in smaller steps.

Output Impedance: The output impedance of a sensor Z_{out} is important to properly interface the sensor with other electronic circuits. The impedance is connected either in parallel with the input impedance (Z_{in}) of the circuit (Thevenin equivalent) or in series (Norton equivalent—see Figure 4.4). To minimize output signal distortions, the sensor-generating current (Norton equivalent) should have output impedance as high as possible and the circuit's input impedance should be very low. For the Thevenin equivalent case, a sensor should have a low Z_{in} and the circuit should have a high Z_{in}.

Excitation: This is the electrical signal required for active transducer operation. Excitation is specified as a range of voltage and/or current. Variations in the excitation can change the transducer's transfer function and cause output errors.

Dynamic Range: One of the subtlest characteristics of sensors, dynamic range occurs when the input varies, and a sensor's response does not really follow with the expected fidelity. This happens because the sensor and its coupling with the stimulus source cannot always respond instantly. Therefore, the response may be time dependent on the dynamic range. Among the dynamic factors in sensors are warm-up time, frequency response, speed response, time constant, lower cutoff frequency, phase shift, and resonant frequency.

Reliability: This is the ability of a sensor to perform a required function under stated conditions for a stated period of time.

4.2. PHYSICAL PRINCIPLES OF SENSING

A sensor is a device that receives a signal or stimulus and responds with an electrical signal. A transducer is basically a converter of one type of energy to another. Sensors and their associated circuits are used to measure various physical properties such as temperature, force pressure, flow position, light intensity, and electric and magnetic fields. The sensor output must be conditioned and processed to provide a valuable measurement. Therefore, sensors must have their output signal conditioned with signal conditioners (e.g., amplifiers) and several analog or digital signal processing circuits.

Sensors can be classified in several ways. From a signal-containing point of view, they can be classified as either active or passive. An active sensor requires an external source for excitation. Resistor-based sensors such as thermistors, resistance temperature detectors (RTDs), and strain gauges are examples of active sensors, because they all require an active current (from an external current source) to go through them and the corresponding voltage must be measured to determine the resistance value. An alternative would be to place the sensor in a bridge circuit in which an external voltage is required. A passive (or self-generating) sensor generates its own electrical output signal without the need of an external voltage or current. Examples of passive sensors are thermocouples and photodiodes, which generate thermoelectric voltages and photocurrents.

The full-scale outputs of most active and passive sensors are relatively small voltages, currents, or resistance changes, and this requires their outputs to be properly conditioned before any further analog or digital processing can occur. These circuits are known as signal-conditioning circuits. Examples of such circuits are amplification, level translation, galvanic isolation, impedance transformation, linearization, and filtering. However, the performance of the

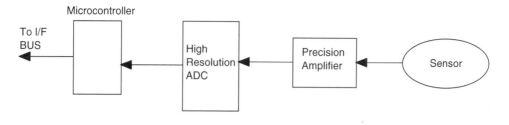

FIGURE 4.5 Data processing in sensors.

sensor will mainly depend on the electrical character of the sensor and its output. Accurate characterization of the sensors in terms of qualities such as sensitivity, voltage and current levels, linearity, impedances, gain, offset, drift, time constants, electrical ratings, and stray impedances can make the difference between substandard and successful application of the sensor, especially in cases where high resolution is desired.

Most sensor outputs are nonlinear with respect to the stimulus, and their outputs must be linearized to yield correct measurements. Analog techniques can be used to perform this function; however, the recent introduction of high-performance analog-to-digital converters (ADCs) enables linearization to be done more effectively in software and eliminates the need for calibration.

Digital techniques are becoming more and more popular in processing sensor outputs in data acquisition, process control, and measurement. By including A–D conversion and the microcontroller programmability on the sensor itself, a "smart sensor" can be implemented with self-contained calibration and linearization features. The basic building blocks of a "smart sensor" are shown in Figure 4.5.

4.3. SENSOR INTERFACING

Resistive elements are some of the most common sensors. Resistive elements can be made sensitive to temperature, strain (by pressure or flexing), and light. Many complex phenomena can be measured; such as fluid flow (by sensing the temperature difference between two calibrated resistances) and dew-point humidity (by measuring two different temperature points).

Bridges offer a good method for measuring small resistance changes accurately. The basic Wheatstone bridge consists of four resistors, as shown in Figure 4.6, where

$$V_0 = \frac{R_1}{R_1 + R_2} V_B - \frac{R_2}{R_2 + R_3} V_B \quad (4.2)$$

and at balance

$$V_0 = 0 \quad \text{if} \quad \frac{R_1}{R_4} = \frac{R_2}{R_3} \quad (4.3)$$

The detector measures the difference between the outputs of two voltage dividers connected across the excitation. A bridge measures resistance indirectly

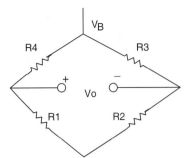

FIGURE 4.6 Basic Wheatstone bridge.

by comparison with a similar resistance. The two principal ways of operating a bridge are as a null detector or as a device that reads differences directly as voltages.

For the majority of sensor applications employing bridges, the deviation of one or more resistors in a bridge from an initial value is measured as an indication of the magnitude in the measured variable. In this case, the output voltage is an indication of the resistance change. Due to very small common resistance changes, the output voltage change may be as small as tens of millivolts. In Figure 4.7, we observe four commonly used bridges suitable for sensor applications and the equations relating the bridge output voltages to the excitation voltages and the bridge resistance values (V_B is assumed to be constant).

The single-element-varying bridge (Figure 4.7a) is most suitable for temperature measurements using RTDs or thermistors. This configuration can also be used for strain gauges. All the resistances are nominal, except for one (the sensor), which varies by an amount ΔR and is not linear.

Linearization of the output of a bridge circuit requires two or more op-amps, as shown in Figure 4.8. The current through each leg of the bridge remains constant ($I_B/2$) as the resistance changes. Therefore the output is a linear function of ΔR. An instrumentation amplifier provides the additional gain.

4.4. DRIVING BRIDGES

Wire resistance and noise pickup are the greatest problems associated with remotely located bridges. For this reason, most bridges are driven using Kelvin or four-wire sensing. The sensing lines go to high-impedance op-amp inputs; thus there is very little error due to the bias-current-induced voltage drop across their lead resistance. The op-amps maintain the required excitation voltage in order to make the measured voltage between the sense leads always equal to V_B.

The second configuration (Figure 4.7b) consists of both elements being in the same direction, such as two identical strain gauges mounted adjacent to each other. The nonlinearity is the same as that of a single-element-varying bridge but with twice the gain. This two-element-varying bridge is commonly found in pressure sensors and flow meter systems. In the third configuration (Figure 4.7c), two identical elements vary in opposite directions. This could correspond to two identical strain gauges: one mounted on top of a flexing

DRIVING BRIDGES

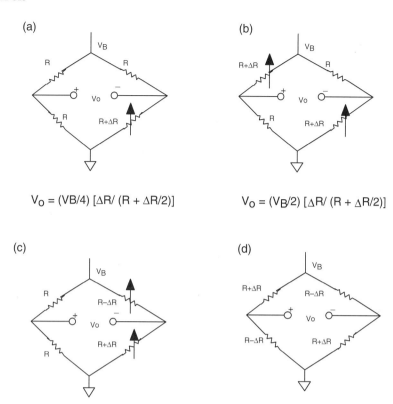

FIGURE 4.7 (a) through (d) are four common bridges for sensor applications.

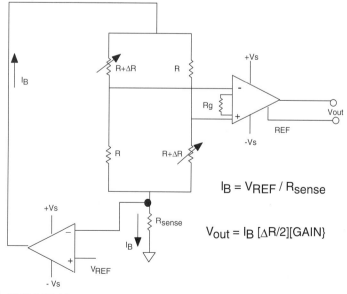

FIGURE 4.8 Linearization of bridges using op-amps.

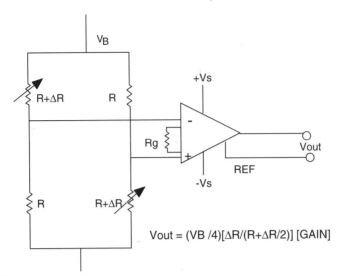

FIGURE 4.9 Single precision op-amp for sensors.

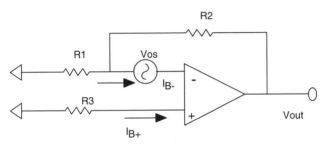

FIGURE 4.10 The effect of input biasing in op-amps.

surface, and one on the bottom. This configuration is linear and has twice the gain of the single-element configuration. The all-element-varying bridge produces the most signal for a given resistance change and is inherently linear.

The output of a single-element-varying bridge can be amplified by a single precision op-amp, connected as shown in Figure 4.9. This efficient circuit provides good gain accuracy (set by resistor Rg). Excellent common mode rejection can be obtained with this amplifier circuit. The output is nonlinear but this can be corrected in the software (by digitizing first the signal using ADCs followed by a microcontroller).

Input bias currents also contribute to the offset error, as shown in the generalized model of Figure 4.10, where noise gain $= 1 + R_2/R_1$ and gain $= -R_2/R_1$.

$$\text{offset to op-amp output} = V_{os}\left[1 + \frac{R_2}{R_1}\right] + I_{B+} \cdot R_3 \left[1 + \frac{R_2}{R_1}\right] - I_{B-} \cdot R_2 \tag{4.4}$$

$$\text{offset to op-amp input} = V_{os} + I_{B+} \cdot R_3 - I_{B-} \left[\frac{R_1 R_2}{R_1 + R_2}\right] \tag{4.5}$$

DRIVING BRIDGES

For bias current cancellation,

$$\text{offset to op-amp input} = V_{os} \quad \text{if} \quad I_{B+} = I_B \quad \text{and} \quad R_3 = \frac{R_1 R_2}{R_1 + R_2}$$

To maintain accuracy a precision amplifier's dc open-loop gain, A_{OL} should be high. This can be observed by examining the equation for closed-loop gain

$$\text{closed-loop gain} = A_{CL} = \frac{NG}{1 + (NG/A_{OL})} \quad (4.6)$$

Noise gain (NG) is the gain seen by a small voltage source in series with op-amp input and is also the amplifier signal gain in the noninverting mode. For finite values of A_{OL} (open-loop gain), there is a closed-loop gain error given by the equation.

$$\% \text{ Gain Error} = \frac{NG}{NG + A_{VOL}} \times 100\% \approx \frac{NG}{A_{VOL}} \times 100\% \quad \text{for } NG \ll A_{VOL}$$

Although Kelvin sensing eliminates errors due to voltage drops in the wiring resistance, the drive voltages must still be highly stable since they directly affect the bridge output voltage (see Figure 4.11). In addition, the op-amps must have low effect, low drift, and low noise.

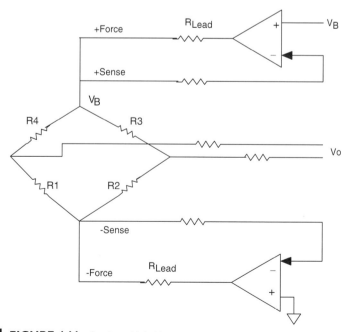

FIGURE 4.11 Sensing with bridges.

4.5. SIGNAL-CONDITIONING AMPLIFIERS

The challenge of selecting the right amplifier for a particular signal-conditioning application is complicated by the wide selection of processes, architectures, and such amplifiers. The following are characteristics that should be looked for when choosing signal conditioning amplifiers: (1) input offset voltage < 100 μV; (2) input offset voltage drift, 1 μV/°C; (3) input bias current < 2 nA; (4) input offset current < 2 nA; (5) dc open-loop gain 71,000,000; (6) unity gain bandwidth product between 500 kHz and 5 MHz; (7) $1/f$ noise < 1 μV p-p; (8) wideband noise < 10 nV/$\sqrt{\text{Hz}}$; and (9) CMR, PSR > 100 dB.

Input offset voltage is one of the largest error sources for precision amplifier circuit design. It is a systemic error that can usually be addressed by using a manual offset null-trim-only system calibration technique using a microcontroller. However, today's amplifiers are very low-offset amplifiers and it is possible to eliminate the need for manual trims.

$$\% \text{ gain error} \cong \frac{\text{NG}}{A_{\text{OL}}} \times 100\% \quad \text{for NG} \ll A_{\text{OL}} \qquad (4.7)$$

Changes in the output voltage and the output loading are the most common causes of changes in the open-loop gain of op-amps. A change in open-loop gain with signal levels produces nonlinearities in the closed-loop gain transfer function, which cannot be removed during system calibration. The severity of the nonlinearity varies widely from device to device, and is generally not specified on the data sheet. The minimum A_{OL} is always specified and choosing an op-amp with a high A_{OL} will minimize the probability of gain nonlinearity errors. One of the most common causes of linearity is thermal feedback.

4.5.1. Noise

There are basically three noise sources in an op-amp circuit. They are the voltage noise of the op-amp, the current noise of the op-amp, and the Johnson noise of the resistances in the circuit. Op-amp noise has two components: "white" noise of medium frequencies and the low-frequency noise. The low-frequency noise is generally known as $1/f$ noise. The frequency at which the $1/f$ noise spectral density equals the white noise is known as the $1/f$ corner frequency F_c and is a figure of merit for an op-amp.

The equation for the total root mean square (rms) noise $V_{n,\text{rms}}$ in the bandwidth $B = F_H - F_L$ is given in the equation

$$V_{n,\text{rms}}(F_H, F_L) = V(H)\sqrt{F_c \ln\left[\frac{F_H}{F_L}\right] + (F_H - F_L)} \qquad (4.8)$$

where $V(H)$ is the noise spectral density in the white noise region, and F_c is the $1/f$ corner frequency. At higher frequencies, the preceding equation is

$$V_{n,\text{rms}}(H) \cong V_{\text{nw}}\sqrt{F_H - F_L}$$

SIGNAL-CONDITIONING AMPLIFIERS

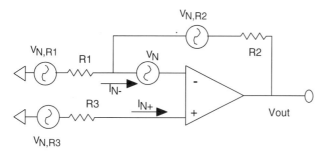

$V_{N,R1} = \{4KTR1\}^{1/2}$ $V_{N,R2} = \{4KTR2\}^{1/2}$ $V_{N,R3} = \{4KTR3\}^{1/2}$

FIGURE 4.12 Noise source representations in an op-amp.

and if $F_H \gg F_L$,

$$V_{n,\text{rms}} \approx V_{n\omega}\sqrt{F_c \ln\left(\frac{F_H}{F_L}\right)} \tag{4.9}$$

At very low frequencies when operating exclusively in the $1/f$ region $F_c \gg (F_H - F_L)$, and the expression for the rms noise reduces to

$$V_{n,\text{rms}}(H) \approx V_{nw}\sqrt{F_c \ln\left[\frac{F_H}{F_L}\right]} \tag{4.10}$$

A generalized noise model for an op-amp is shown in Figure 4.12. All uncorrelated noise sources add as a root-sum-of-squares manner, that is, noise voltages V_1, V_2, and V_3 give a result of

$$\sqrt{V_1^2 + V_2^2 + V_3^2}$$

Thus, any noise voltage that is more than four or five times any of the others is dominant, and the others can generally be ignored.

The gains are given by

$$\text{gain} = \frac{-R_2}{R_1} \qquad \text{NG} = 1 + \frac{R_1}{R_2}$$

the offset referred to the input noise is given by Eq. (4.11)

$$\text{offset} = \sqrt{B(H)}\left(V_N^2 + 4KTR_3 + 4KTR_1\left[\frac{R_2}{R_1+R_2}\right]^2 + I_N^2 R_3^2 \right.$$

$$\left. + I_{N-}^2\left[\frac{R_1 R_2}{R_1+R_2}\right] + 4KTR_2\left[\frac{R_1}{R_1+R_2}\right]^2\right)^{\frac{1}{2}} \tag{4.11}$$

Johnson noise is associated with the term $\sqrt{4kTBR}$, where K is Boltzmann's constant (11.38×10^{-25} J/K), T is the absolute temperature, B is the bandwidth in hertz, and R is the resistance in ohms. An easy rule-of-thumb to remember is that a 1000-Ω resistor generates a Johnson noise of $4nV/\sqrt{Hz}$ at 25°C. Bipolar input op-amps have lower noise margins than junction FET (JFET) input op-amps. For bipolar or JFET input devices, where all the bias current flows into the input function, the current noise is simply the shot noise of the bias current.

4.6. INSTRUMENTATION AMPLIFIERS

An instrumentation amplifier is basically a closed-loop gain block, which is single-ended with respect to a reference terminal (see Figure 4.13). The input impedances are balanced and have high values, typically $10^9 \Omega$. Unlike an op-amp, which has its closed-loop gain determined by external resistors connected between inverting input and its output, an instrumentation amplifier employs an internal feedback resistor network that is isolated from its signal input terminals. With the input signal applied across the two differential inputs, gain is either present internally or is user set via pins or an external gain resistor, which is isolated from the signal inputs.

The instrumentation amplifier must be capable of amplifying microvolt-level signals, while simultaneously rejecting volts of common mode signals of its inputs. This requires that instrumentation amplifiers have very high common mode rejection (CMRR) on the order of 70 to 100 dB.

Two op-amp instrumentation amplifiers are often used (see Figure 4.14). Dual IC op-amps are used in most cases for good matching. The circuit gain can be trimmed with an external resistor, Rg. The input impedance is high, allowing the impedance of the signal sources to be high and unbalanced.

When dual supplies are used, V_{ref} is normally connected directly to ground. In single-supply applications, V_{ref} is usually connected to a low-impedance voltage source equal to one-half the supply voltage. The best bipolar amplifiers offer offset voltage of 10 μV, and 0.1-μV/°C and drift offset voltages less than 5 μV, with practically no measurable effect are obtainable with choppers.

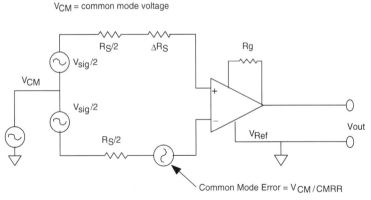

FIGURE 4.13 Instrumentation amplifier.

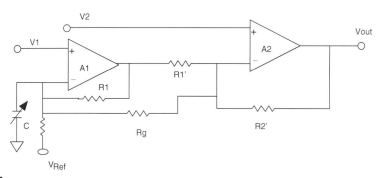

FIGURE 4.14 Dual op-amp instrumentation amplifier.

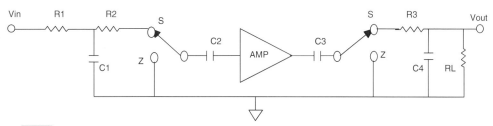

FIGURE 4.15 Chopper stabilized design.

The basic chopper design is shown in Figure 4.15. When the switches are in the "Z" (autozero) position, capacitors C2 and C3 are charged to the amplifier input–output offset voltage, respectively. When the switches are in the "S" (sample) position, V_{in} is connected to V_{out} through the path comprised of $R1$, $R2$, $C2$, the amplifier, $C3$, and $R3$. The chopping frequency is between a few hundred hertz and several kilohertz. Due to the sampling, the input frequency must be much less than one-half the chopping frequency to prevent errors due to aliasing. The $R1/C1$ combination serves as an antialiasing filter. It is also important to state that after a steady condition is reached, there is only a minimal amount of charge transferred during the switching cycles. The output capacitor, $C4$, and the load, R_L, must be chosen such that there is minimal V_{out} droop during the autozero cycle.

One major disadvantage of the two-amp design is that common mode stage input range must be traded off for gain. The amplifier $A1$ must amplify the signal at $V1$ by $1 + R1/R2$ if $R1 \gg R2$. $A1$ will saturate if the common mode signal is too high, leaving very little room to amplify the wanted differential signal.

For truly balanced high-impedance inputs, three op-amps may be connected to form the instrumentation amplifier shown in Figure 4.16. The gain of the amplifier is set by the resistor, Rg, which may be internal, external, or programmable. For this case the common mode rejection ratio (CMRR) depends on the ratio matching $R3/R2$ to $R3'/R2'$. Furthermore, common mode signals are amplified by a factor of 1 only regardless of gain. Thus, CMRR will increase in direct proportion to the gain.

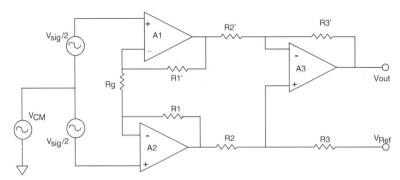

FIGURE 4.16 Three op-amp instrumentation amplifier.

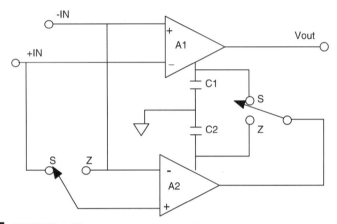

FIGURE 4.17 Chopper stabilized amplifier.

4.7. CHOPPER-STABILIZED AMPLIFIERS

If we want the lowest offset and lowest drift performance, chopper-stabilized amplifiers may be the only solution. The chopper-stabilized architecture shown in Figure 4.17 is most often used in chopper amplifier implementations. In this circuit, $A1$ is the main amplifier implementation. In this circuit, $A1$ is the main amplifier and $A2$ is the nulling amplifier. In the sample mode (switch in "S" position), the nulling amplifier $A2$, monitors the input offset voltage of $A1$ and drives its output to zero by applying a correcting voltage at $A1$'s null pin. Also $A2$ has an input offset voltage: it must correct its own zero error before attempting to null $A1$'s offset. This can be achieved in the autozero mode (switches in "Z" position) by momentarily disconnecting $A2$ from $A1$, shorting its inputs together, and coupling its output to its own null pin. During autozero mode the correction voltage for $A1$ is held by $C1$. Similarly, $C2$ holds the correction voltage for $A2$ during the sample mode.

The switching action does produce small transients at the chopping frequency, which can mix with the input signal frequency and produce in-band distortions.

4.8. ISOLATION AMPLIFIER

There are applications where a sensor has no direct electrical connection with the system to which it supplies the signal data. For these cases, an isolation amplifier is used to avoid the possibility of dangerous voltages or currents from one-half of the system doing damage in the other, or to break an intractable ground loop. Examples of applications include the need to prevent the ignition of explosive gases by sparks at sensors and the protection from electric shock of patients whose ECG, EEG, or EMG is being monitored. The ECG is important since it requires protection in both directions: the patient must be protected from accidental electric shock, but if the patient's heart should stop, the ECG machine must also be protected from the more than 7.5 kV applied to the patient by the defibrillator, which will be used to start the patient's heart again.

The most common type of isolation amplifier uses transformers that use magnetic fields. Other amplifiers use small high-voltage capacitors. Optoisolators provide isolation by using light. Difference isolators have different performance; transformers have an analog accuracy of 12 to 16 bits and bandwidth up to several hundred kilohertz, but their maximum voltage rating rarely exceeds 10 kV and is usually much lower. Capacitively coupled isolation amplifiers have lower accuracy, perhaps 12 bits maximum, lower bandwidths, and lower voltage rating. Optoisolators are fast and cheap and can be made with very high voltage ratings (4–7 kV), but they have poor analog domain linearity and must be provided one—a transformer can do well here. The isolation amplifier has an input circuit that is galvanically isolated from the power supply and the output circuit. There is also minimal capacitance between the input and the rest of the device. An example of a three-part isolation amplifier is shown in Figure 4.18.

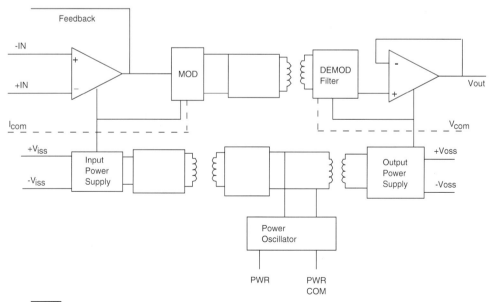

FIGURE 4.18 Example of isolation amplifier.

4.9. STRAIN, FORCE, PRESSURE, AND FLOW SENSORS

The most typical sensors used in measuring forces include the resistance strain gauge, the semiconductor strain gauge, and piezoelectric transducers. The strain gauge measures force indirectly by measuring the deflection it produces in a calibrated carrier. Pressure can also be converted into a force using an appropriate transducer, which means that strain gauge techniques can be used to measure pressure. The flow rates of a liquid can be measured using differential pressure measurements, which also make use of strain gauge technology.

The resistance strain gauge is a resistive element that changes in length (and its resistance) as the force applied to the base on which the resistive element is mounted causes stretching or compression. An unbounded strain gauge consists of a wire stretched between two points, as shown in Figure 4.19. The force acting on the wire (area $= A$, length $= L$, resistivity $= \rho$) will cause the wire to either elongate or shorten and this will cause the resistance to either increase or decrease according to the equation

$$R = \frac{\rho L}{A}$$
$$\frac{\Delta R}{R} = \frac{\rho}{A} \frac{\Delta L}{L} \tag{4.12}$$

The quantity $\Delta L/L$ is a measure of the force applied to the wire and is expressed in microstrains $(1\,\mu\varepsilon = 10^{-6}\,\text{cm/cm})$ which is the same as parts per million.

Bonded wire strain is a thin wire or conductive film arranged in a coplanar pattern and cemented to a base. Lead wires are attached to the base and brought out for interconnection. Bonded devices are much more practical. Another variation of the bonded type is the foil-type gauge produced by photoetching. Foil sensing elements have large ratios of surface area to cross-sectional area and are stabler under extremes of temperature and prolonged loading (Figure 4.20).

Semiconductor strain gauges make use of the piezoresistive effect in certain semiconductor materials such as silicon and germanium to obtain greater sensitivity and higher-level outputs. Semiconductor strain gauges are very temperature sensitive and difficult to compensate and their charge in resistance with a force that is nonlinear. They are not widely used unless for very sensitive applications where temperature variations are small.

Piezoelectric force transducers are used when the forces to be measured are dynamic (usually for several milliseconds). They work from the effect that changes in charge are produced in certain materials when they are subjected

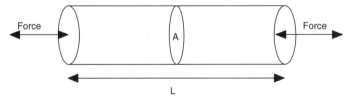

FIGURE 4.19 Strain forces on a wire.

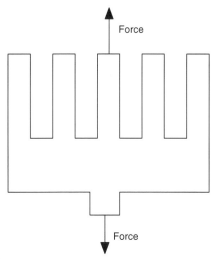

FIGURE 4.20 Planar foil strain gauge.

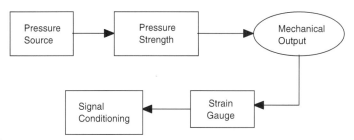

FIGURE 4.21 Pressure transducer overview.

to physical stress. Piezoelectric devices produce large output voltage in instruments such as accelerometers for vibration studies. Their output impedance is high, and charge amplification with low-input capacitance is required for signal conditioning.

Pressure in liquids and gases is measured electrically by a variety of pressure transducers (see Figure 4.21). A variety of mechanical converters (e.g., diaphragms, capsules, bellows, manometer tubes) are used to measure pressure by measuring the associated length of the displacement, thus measuring pressure by the motion produced.

Flow (volume, laminar, and turbulent) can be measured by taking the differential pressure across two points in the medium, one at a static point and one in the flow stream. The flow rate is obtained by measuring the differential pressure with a pressure transducer, as shown in Figure 4.22a. Differential pressure can also be measured by the venturi effect, as shown in Figure 4.22b.

An example of an all-element varying bridge circuit is shown in Figure 4.23. The full bridge is an integrated unit that can be attached to the surface where the force is going to be measured. To facilitate remote sensing, current excitation is used. In the Analog Devices OP177 servos, a bridge current goes to 10 mA around a reference voltage of 1.235 V. The strain gauge produces an output of

92   4 SENSOR CHARACTERISTICS

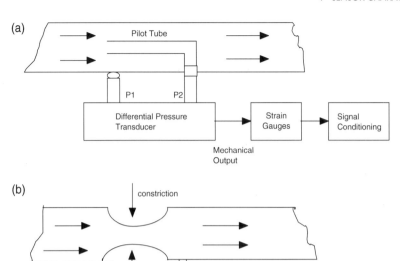

FIGURE 4.22 Flow sensor overview. (a) Non-constriction, (b) constriction.

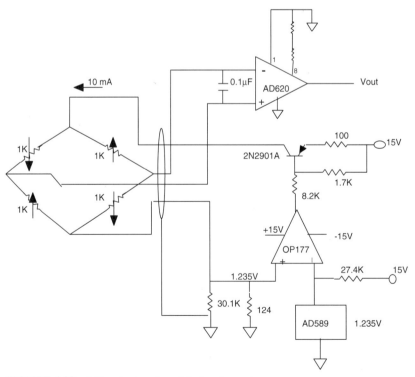

FIGURE 4.23 Bridge circuit and conditioning electronics.

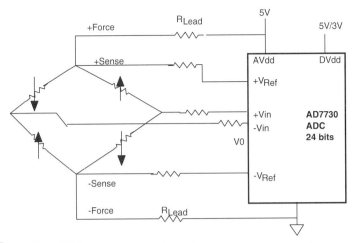

FIGURE 4.24 Bridge circuit with an ADC.

10.25 mV/1000 µε. The signal is then amplified by the Analog Devices AD620 instrumentation amplifier configured for a gain of 100. Full-scale strain voltage can be set by adjusting the 100-Ω gain potentiometer so that for a strain of −35,000 µε, the output reads −3.5 V; and for a strain of 5000 µε, the output reads 5.0 V. The output can be digitized with an ADC with a 10-V full-scale input range.

The Analog Devices AD7730 24-bit sigma delta ADC allows direct conditioning of bridge outputs and requires no interface circuiting. The diagram is shown in Figure 4.24. The entire circuit operates on a single 5-V supply that also functions as the bridge excitation voltage. Variations of the 5-V supply do not affect the accuracy of the measurement. The AD7730 has an internal programmable gain amplifier that allows a full-scale bridge output of ±10 mV to be digitized to 16-bit accuracy. The AD7730 has self- and system calibration features that enable offset and gain errors to be minimized.

4.10. HIGH-IMPEDANCE SENSORS

Many sensors have output impedances greater than several megaohms and the associated signal conditioning circuits must be designed to meet the requirements of low bias current, low noise, and high gain. Examples of high-impedance sensors include photodiode preamplifier, piezoelectric sensors, charge output sensors, and charge-coupled devices.

The equivalent circuit for a photodiode is shown in Figure 4.25. Its medical application is in computed axial tomography (CAT) scanners (x-ray detection) and blood particle analyzers. Its most important parameter is its short-circuit photocurrent (I_{sc}) at a given level from a well-defined light source. The short resistance R_{sh} is usually of the order of 100 MΩ at room temperature. Diode capacitance C_j is a function of junction area and the diode bias voltage (∼50 pF). Photodiodes operate most often in the photovoltaic mode (see Figure 4.26).

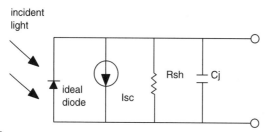

FIGURE 4.25 Equivalent circuit of a photodiode.

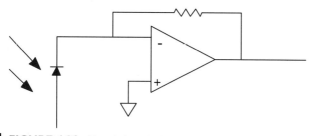

FIGURE 4.26 Typical photodiode mode operation.

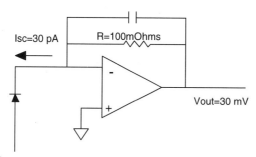

FIGURE 4.27 Current-to-voltage converter using a photodiode.

In the photovoltaic mode, the diode operates in linear mode and the only noise is thermal noise. A convenient way to convert the photodiode current into a usable voltage is to use an op-amp as a current-to-voltage converter. The diode bias is kept at 0 V by the virtual ground of the op-amp and the short-circuit current is converted into a voltage.

The amplifier must be able to detect a diode current of 30 pA (see Figure 4.27). For higher currents, smaller feedback resistors can be used. Since the diode current is measured in terms of picoamperes, a great deal of attention must be paid to potential leakage paths in the actual circuit. The feedback resistor should be thin film of ceramic or glass with glass insulation. The compensation capacitor across the feedback resistor should have a polypropylene or polystyrene dielectric. All connections to the summing junction should be kept short. Guarding techniques can be used to reduce parasitic leakage currents isolating amplifier's input from large voltage gradients across the PC board. Physically, a guard is a low-impedance conductor that surrounds an input line and is raised to the line's voltage.

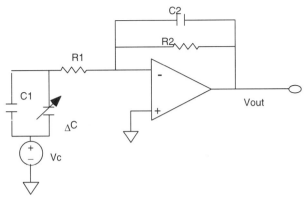

FIGURE 4.28 Inverting charge-sensitive amplifier.

4.11. HIGH-IMPEDANCE CHARGE OUTPUT

High-impedance transducers such as piezoelectric sensors, hydrophones, and some accelerometers require an amplifier that will convert a transfer of charge into a change of voltage. Because of the high dc output impedance of these devices, appropriate buffers are required in basic circuits for an inverting charge-sensitive amplifier, as shown in Figure 4.28. There are basically two types of charge transducers: capacitive and charge emitting. For the capacitive transducer, the voltage across the capacitor (V_C) is held constant. The change in capacitance ΔC produces a change in charge $\Delta Q = \Delta C V_C$. This charge is transferred to the op-amp output as a voltage $\Delta V_{\text{out}} = -\Delta Q/C2 = -\Delta C V_C/C2$.

For capacitive sensors $\Delta V_{\text{out}} = (-V_C \Delta C)/(C2)$ and for charge-emitting sensors $\Delta V_{\text{out}} = \Delta Q/C2$. The upper cutoff frequency is given by $f_2 = 1/(2\pi R2\, C2)$ and the lower by $f_1 = 1/(2\pi R1 C1)$.

Figure 4.29 shows two ways to buffer and amplify the output of a charge output transducer. Both require using an amplifier that has a very high input impedance, such as Analog Devices AD745. The AD745 provides both low voltage and low current noise, and this combination makes this device particularly suitable in applications requiring very high charge sensitivity.

4.12. CHARGE-COUPLED DEVICE SENSORS

The charge-coupled device (CCD) and the contact image sensor (CIS) are widely used in imaging systems. A block diagram of an imaging system is shown in Figure 4.30. The imaging sensor CCD is exposed to the image or picture. After exposure, the output of the CCD undergoes some analog signal processing and then the signal is digitized by an ADC. Most of the processing of the image is done by digital signal processors.

The building blocks of a CCD are individual sensing elements (Figure 4.31). A sensing element consists of a photodiode or photocapacitor that outputs a

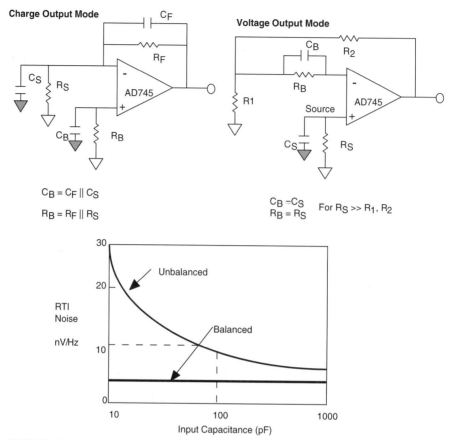

FIGURE 4.29 Balancing source impedances minimizes the effects of bias currents and reduces input noise.

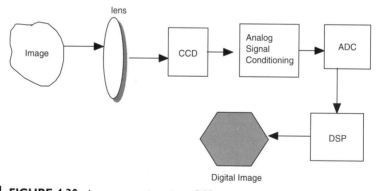

FIGURE 4.30 Image processing using a CCD.

charge (electrons) in proportion to the incoming light. The charge is accumulated during the exposure time and then the charge is transferred to the CCD shift register to be sent to the output of the device. The amount of accumulated charge depends on the light level, the integration time, and the quantum

CHARGE-COUPLED DEVICE SENSORS

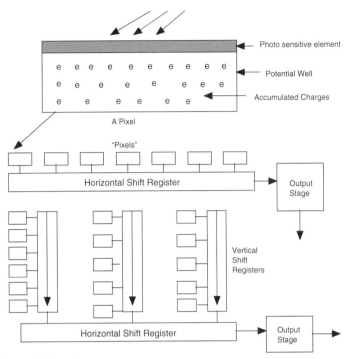

FIGURE 4.31 The CCD sensor.

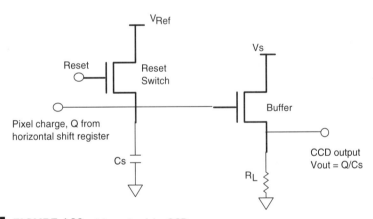

FIGURE 4.32 Schematic of the CCD output stage.

efficiency of the photosensitive element. A small amount of charge always accumulates even without the presence of light. This is called the dark current and must be taken into account during the signal processing.

A typical CCD output stage is shown in Figure 4.32. The output stage of the CCD converts the charge in each pixel to a voltage via the sensing capacitor C_s. At the beginning of each pixel period, the voltage on C_s is reset to the reference level V_{Ref}, causing the switch to reset. The amount of light captured by each pixel is resolved by the difference between the reference and the video level ΔV.

The CCD charges can be as low as 10 electrons and a typical CCD output has a sensitivity of 0.6 μV/electron. Most CCDs have a saturation voltage of about 500 mV to 1 V for area sensors and 2 to 4 V for linear sensors. The dark current (DC) level of the waveform is between 3 and 7 V.

The output of the CCD is processed by external conditioning circuits. Therefore, the CCD output must be clamped before being digitized by the ADC. CCD output voltages are small and often buried in noise. The largest noise source is the thermal noise in the resistance of the FET reset switch. This noise range is of the order of 100 to 300 electrons rms ($\approx$ 60 to 180 mV rms) and is given by the expression

$$\text{thermal noise} = \sqrt{4KT \cdot \text{BW} \cdot R_{on}} \qquad (4.13)$$

and its bandwidth is given by noise BW = $1/(4R_{on}C_s)$, where R_{on} is the "on" resistance of the reset switch. When the reset switch opens, the KT/C_s is stored on the C_s and remains constant until the next reset interval.

4.13. POSITION AND MOTION SENSORS

Modern linear and digital integrated circuits are used in the areas of position and motion sensing. An example of a linear integrated sensor is the linear variable differential transformer.

4.13.1. Linear Variable Differential Transformers

The linear variable differential transformer (LVDT) is a reliable method for measuring linear distance. The LVDT method positions an electrical sensor where the output is proportional to the position of a movable magnetic core. The core moves linearly inside a transformer consisting of a center primary coil and two outer secondary coils wound up in a cylindrical form. The primary winding is excited by an ac voltage source (of several kilohertz), where secondary voltages are induced by varying the position of the magnetic core within the assembly. The core is usually threaded to facilitate attachment to a nonferromagnetic rod, which in turn is attached to the object whose movement or displacement is being measured.

When the core is centered, the voltages in the two secondary windings oppose each other, and the net voltage is zero. As the core is moved off center, the voltage in the secondary winding in the direction in which the core moves increases while the opposite voltage decreases. The LVDT offers good accuracy, linearity, sensitivity, and infinite resolution. For LVDTs, a typical excitation voltage range is from 100 μm to 25 cm and the excitation voltages range from 1 to 24 V with frequencies ranging from 50 Hz to 20 kHz.

A signal conditional circuit for LVDT is shown in Figure 4.33, where both positive and negative variations about the center position can be measured in an industry standard. An LVDT signal conditioner developed by Analog Devices is shown in Figure 4.34. Note that the AD698 operates from a four-wire LVDT and uses synchronization demodulation.

POSITION AND MOTION SENSORS

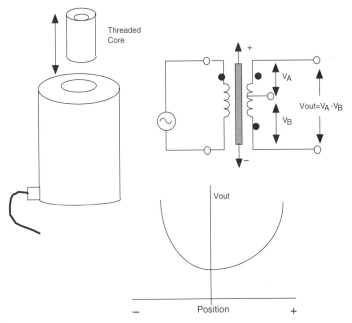

FIGURE 4.33 Physical description of the LVDT.

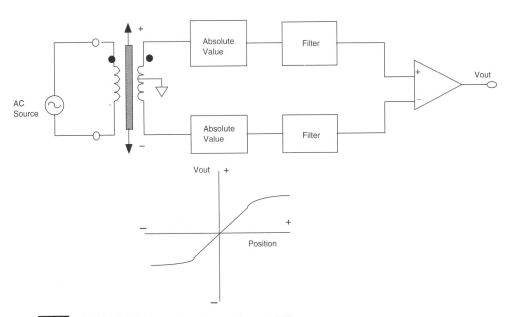

FIGURE 4.34 Signal conditioning for an LVDT.

The *A* and *B* signal processors each consist of an absolute value function and a filter. The *A* output is then divided by the *B* output to produce a final output that is ratiometric and independent of the excitation voltage amplitude. The AD698 can also be used with a half-bridge (similar to an autotransformer) LVDT, as shown in Figure 4.35. In this arrangement, the secondary voltage

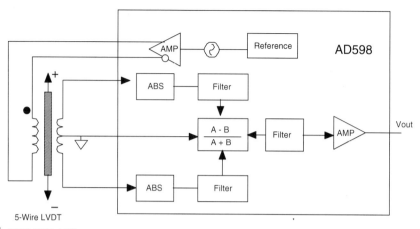

FIGURE 4.35 Example of signal conditioning circuit.

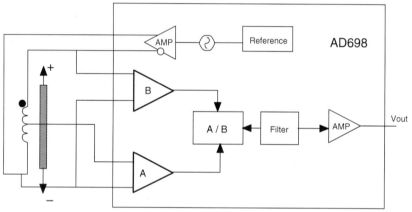

FIGURE 4.36 Half-bridge LVDT configuration.

is applied to the B processor, while the center-tap voltage is applied to the A processor. The half-bridge LVDT does not produce a null voltage, and the A/B ratio represents the range-of-travel of the core (Figure 4.36).

4.13.2. Hall Effect Magnetic Sensors

If a current flows in a conductor (or semiconductor) and there is a magnetic field B present that is perpendicular to the current flow, the combination of the current in the presence of a magnetic filed will generate a voltage that is perpendicular to both, as shown in Figure 4.37. This phenomenon is called the Hall effect. The voltage V_H is known as the Hall voltage.

The Hall effect can be used to measure the magnetic field, but its most important application is as a motion sensor where a fixed Hall sensor and a small magnet attached to a moving part can replace a CAM and contacts with a great improvement in reliability. Because V_H is proportional to magnetic field

POSITION AND MOTION SENSORS

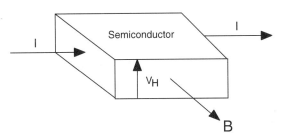

FIGURE 4.37 Physical principles of the Hall effect.

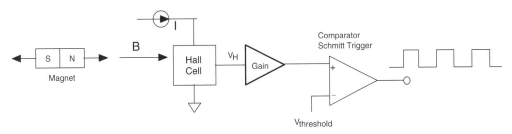

FIGURE 4.38 Signal processing in Hall effect sensors.

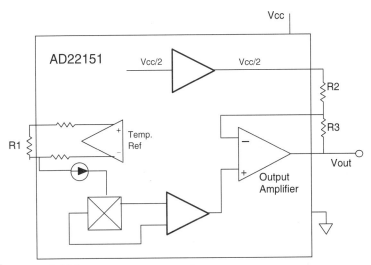

FIGURE 4.39 Hall effect sensor on a chip.

and not to rate of change of magnetic field, the Hall effect provides a more reliable low-speed sensor than an inductive pickup. The simple motion detector can be seen in Figure 4.38. It responds to small changes in the field and the comparator has built-in hysteresis to prevent oscillations. The Analog Devices AD22151 (Figure 4.39) is a linear magnetic field sensor in which the output voltage is directly proportional to a magnetic field applied perpendicular to the sensor's top surface. The AD22151 combines bulk Hall cell technology and conditioning circuitry to minimize temperature-related drifts in silicon Hall cells.

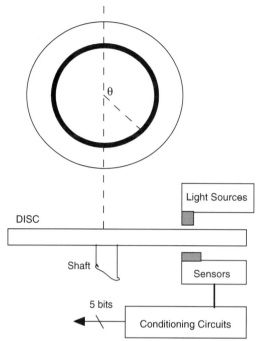

FIGURE 4.40 Physical illustration of an optical encoder.

4.13.3. Optical Encoders

One of the most common position measuring sensors is the optical encoder (Figure 4.40). An incremental optical encoder is a disk divided into sectors that are either transparent or opaque. A light source is positioned on one side of the disk and a light sensor on the other side. When the disk rotates, the output from the detector switches on and off in an alternating manner, depending on whether the sector appearing between the light source and the detector is transparent or opaque. Therefore, the encoder produces a stream of square-wave pulses that indicate the angular position of the shaft. The number of opaque and transparent sectors per disk range from 100 to 65,000.

Most incremental encoders feature a second light source and sense the angle between the main source and the sensor to indicate the direction of rotation. Many other encoders also have a third light source and detector to sense a once-per-revolution marker. A potentially serious disadvantage is that incremental encoders require an external counter to determine absolute angles within a given rotation.

4.13.4. Accelerometers

Accelerometers find great usage in many applications including medical ones (e.g., patient monitors). Micromachining techniques enable accelerometers to be manufactured on a CMOS processor. A significant advantage of this type of accelerometer is that dc acceleration can be measured. The basic unit-cell

POSITION AND MOTION SENSORS

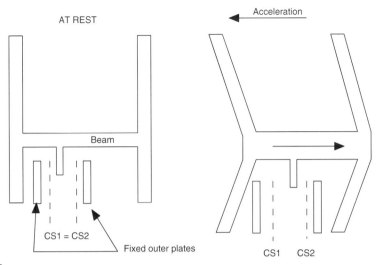

FIGURE 4.41 Accelerometer etched on silicon.

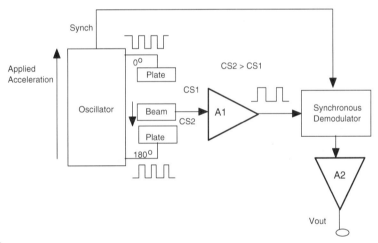

FIGURE 4.42 Conditioning electronics in accelerometers.

building block for these accelerometers is shown in Figure 4.41. The micromachined sensor elements are obtained by depositing polysilicon on an oxide layer that is then etched away, leaving the suspended sensor element.

The actual sensor has tens of unit cells for sensing acceleration (the figure shows one cell). The foundation of the sensor is the differential capacitors (CS1 and CS2), which are formed by a center plate that is part of the moving beam and the two fixed outer plates. The two capacitors are equal at rest, but when acceleration is applied, the mass of the beam causes it to move closer to one of the fixed plates while moving away from the other. The conditioning electronics are shown in Figure 4.42.

The sensor's fixed capacitor plates are driven differentially by a 1-MHz square wave and the two square-wave amplitudes are of the same magnitude but 180° out of phase. At rest, the magnitude of the two capacitors is the same

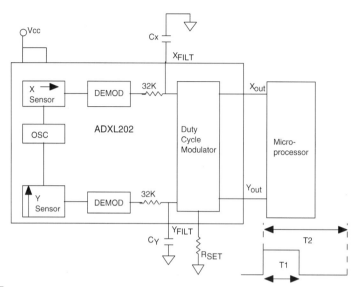

FIGURE 4.43 Accelerometer block diagram on a chip.

and therefore the voltage output at the center plate attached to the movable beam is zero. When the beam starts to move, a mismatch in the capacitances produces an output signal at the center plate. The output amplitude will increase with the acceleration experienced by the sensor. The center plate is buffered by $A1$ and is applied to a synchronous demodulator.

Figure 4.43 shows a simplified block diagram of the Analog Devices ADXL202 dual-axis $\pm 2g$ accelerometer. The output is a pulse whose duty cycle contains the acceleration information. Standard low-cost microcontrollers have timers that can be easily used to measure T_1 and T_2 intervals; the acceleration is calculated in terms of g.

4.14. TEMPERATURE SENSORS

Knowledge of a system's temperature is of great importance for medical devices since temperature monitoring is necessary to monitor the health of electronic devices and avoid dangerous power interruptions.

Except for IC sensors, all temperature sensors have nonlinear transfer functions. However, this is really not an issue since sensor output can be digitized directly by high-resolution ADCs; hence, linearization and calibration are performed digitally. Modern semiconductor temperature sensors offer high accuracy and high linearity over the operating range of -55 to $150°C$. Internal amplifiers can scale the output to values such as $10\,mV/°C$. These semiconductor temperature sensors can be integrated into multifunction ICs and perform other hardware monitoring functions. Thermistors are more sensitive than semiconductor temperature sensors but most are nonlinear. Thermocouples are small, rugged, relatively inexpensive, and operate over the widest range of all temperature sensors. They are specially useful for making measurements

TABLE 4.1 Common thermocouples

Junction materials	Typical useful range (°C)	Normal sensitivity (μV/°C)	ANSI designation
Platinum (6%)–rhodium to platinum (30%)–rhodium	38 to 1800	7.7	B
Tungsten (5%)–rhenium to tungsten (26%)–rhenium	0 to 2300	16	C
Chromel–constantan	0 to 982	76	E
Iron–constantan	0 to 760	55	J
Chromel–alumel	−184 to 1260	39	K
Platinum (13%)–rhodium to platinum	0 to 1593	11.7	R
Platinum (10%)–rhodium to platinum	0 to 1538	10.4	S
Copper–constantan	−184 to 400	45	T

TABLE 4.2 Types of temperature sensors

Thermocouple	RTD	Thermistor	Semiconductor
Widest range: −184 to 2300°C	Range: −200 to 850°C	Range: 0 to 100°C	Range: −55 to 150°C
High accuracy and repeatability	Fair linearity	Poor linearity	Linearity: 1°C Accuracy: 1°C
Requires cold junction compensation	Requires excitation	Requires excitation	Requires excitation
Low voltage output	Low cost	High sensitivity	10 mV/K, 20 mV/K, or 1 μA/K typical output

of high temperatures (up to 2300°C) in hostile environments. Thermocouples can produce only a few millivolts of output and this output requires further amplification before processing. Some common thermocouples are shown in Table 4.1. Table 4.2 shows the most common types of temperature sensors. Of the thermocouples shown, type J is the most sensitive, causing the largest output voltage for a given temperature change. On the other hand, type S is the least sensitive.

Modern semiconductor temperature sensors offer high accuracy and linearity. Internal amplifiers can output to convenient values. All semiconductor temperature sensors make use of the relationship between a bipolar junction transistor (BJT) base-emitter voltage and its collector current

$$V_{BE} = \frac{KT}{q} \ln\left(\frac{I_c}{I_s}\right) \quad (4.14)$$

where K is the Boltzmann's constant, T is the absolute temperature in K, q is the charge of an electron, and I_s is a current related to the geometry and the

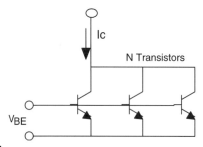

FIGURE 4.44 Basic relationship for semiconductor temperature sensor.

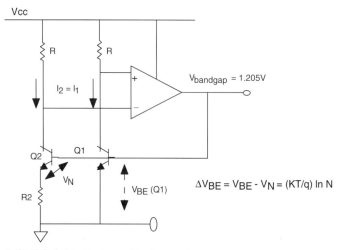

FIGURE 4.45 Brokaw cell implementation.

temperature of the transistor. If we take N transistors identical to the first and allow the total current I_s to be shared equally among them, we obtain a new base-emitter voltage given by the equation (see Figure 4.44).

$$V'_{BE} = \frac{KT}{q} \ln\left(\frac{I_c}{NI_s}\right) \qquad (4.15)$$

If we have equal currents in one BJT and N similar BJTs, then absolute expression for the difference between the two base-emitter voltages is proportional to the absolute temperature and does not contain I_s.

$$\Delta V_{BE} = V_{BE} - V'_{BE} = \frac{KT}{q} \ln\left(\frac{I_c}{I_s}\right) - \frac{KT}{q} \ln\left(\frac{I_c}{NI_s}\right) - \frac{KT}{q} \ln[N] \qquad (4.16)$$

The circuit shown in Figure 4.45 implements this equation and is known as the "Brokaw cell." The voltage $\Delta V_{BE} = \Delta_{BE} - V_n$ appears across resistor $R2$. The emitter current in $Q2$ is therefore $\Delta V_{BE}/R2$. The op-amp servo loops and the resistors R force the same current to flow through $Q1$. The $Q1$ and $Q2$ currents are equal and are summed and flow into resistor $R1$.

The corresponding voltage developed across $R1$ is proportional to absolute temperature and given by

$$V = \frac{2R1(V_{BE} - V_n)}{R2} = 2\frac{R1}{R2}\frac{KT}{q}\ln N \qquad (4.17)$$

REFERENCES

Eamon, Nash. 1998. "A practical review of common mode and instrumentation amplifiers." *Sensor Magazine*, July, 26–33.

Franco, Sergio. 1998. *Design with Operational Amplifiers and Analog Integrated Circuits*, 2nd ed. McGraw-Hill, New York.

Kitchin, Charles, and Lew Counts. 1991. *Instrumentation Amplifier Application Guide*. Analog Devices.

Smith, Lewis, and Dan Sheingold. 1991. "Noise and operational amplifier circuit." *Analog Dialogue*, 25th Anniversary Issue, pp. 19–31.

Stout, D., and M. Kaufman. 1976. *Handbook of Operational Amplifier Circuit Design*. McGraw-Hill, New York.

Practical Design Techniques for Sensor Signal Conditioning. 1999. Analog Devices.

5
DATA ACQUISITION

5.1. INTRODUCTION

Analog data can be converted into digital form and transmitted over short or long distances. In Figure 5.1 we can observe the simplest digitizing system. It shows the analog-to-digital converter, the power inputs, and an analog single input. The outputs are a digital code word.

The maximum rate at which the input signal can vary and still allow the converter to resolve 1 least-significant bit (LSB) of binary output, irrespective of the waveform, is

$$\frac{dV}{dt} = \frac{2^{-n}V_S}{T} \tag{5.1}$$

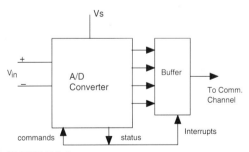

FIGURE 5.1 A simple illustration of an analog-to-digital converter (ADC).

*Thanks are expressed to Analog Devices for the usage of some of their components publications.

where V is the input voltage, n is the number of bits of binary resolution, V_S is the full-scale span, and T is the time between conversions. The maximum rate of changes is thus 1 LBS per conversion period.

If $V = (V_S/2)\sin(2f\pi t)$, then $dV/dt = (V_S/2)2\pi f \cos(2\pi ft)$ and $dV/dt_{\max}$ is equal to the magnitude $(V_S/2)2\pi f$. Thus,

$$\frac{2^{-n}}{T} = \pi f \tag{5.2}$$

and the maximum sine-wave frequency that can be converted with 1-LBS resolution is

$$f = \frac{2^{-n}}{T\pi} \tag{5.3}$$

5.2. SAMPLE AND HOLD CONVERSION

An ADC can be made more efficient by the use of a sample and hold between the input signal and the converter's input (Figure 5.2). Between conversions, the device may acquire and track the input signal. When the conversion is to be made, the S–H is switched to a hold position and remains in that state during the conversion.

To avoid errors due to an insufficient number of samples, the sampling theory says that regularly spaced sampling must occur at the Nyquist rate, which is twice the frequency of the highest frequency signal. In many practical applications, the sampling rate is three or more times the filter cutoff frequency. If analog signals are present at higher frequencies, the sampling process can cause aliasing in the signal passband, which cannot be distinguished from the original signal. Aliasing in essence are spurious low-frequency signals. In the signal passband, they are caused by the difference frequencies produced by the sampling process.

As shown in Figure 5.3, preamplification is often necessary before the sampling of a signal. The preamplifier should have a low-output impedance because

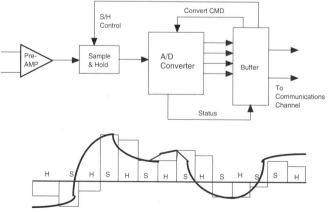

FIGURE 5.2 Sample and hold with an ADC.

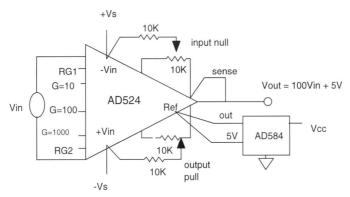

FIGURE 5.3 Signal-conditioning devices in data acquisition.

the inputs of some types of ADCs may have large current pulses, which can load the pre-amp outputs and cause errors.

Signal conditioning is a term that describes a series of analog-to-analog scenarios. For example, sculling of input gains to match the input signal to the converter's full-scale spans using instrumentation amplifiers is an obvious choice. Signal-conditioning devices are available in a variety of packages, as shown in Figure 5.3.

5.3. MULTICHANNEL ACQUISITION

Elements of the acquisition system can be shared by two or more input sources. Although the conventional way to digitize data from many analog channels is by the introduction of a time-sharing process at the analog input by multiplexing the input of a single ADC among the various analog sources; in sequence, the parallel conversion is much more used. The bus structure employed by systems using microprocessors allows the usage of digital multiplexing, with all the devices connected to the bus via three-state switches, enabled selectively by a "chip-select" logic signal from decoders and read–write control signals. The converter's status line can provide interrupt signals, indicating conversion complete data ready.

The parallel-conversion approach provides the advantage that multiple sensors can be strung over a larger area, in essence digitizing the analog signals right at the source and transmitting serial data rather than the original low-level analog signals. The basic multichannel conversion scheme using digital multiplexing is shown in Figure 5.4a. The multichannel conversion scheme using remote ADCs is shown in Figure 5.4b. Furthermore, among the secondary benefits of digitizing sensor signals at their source is the feasibility to manipulate logical operations on the digitized data before they are fed into the computer.

Figure 5.5 shows a basic system for data acquisition. When the conversion is complete, the status signal from the converter causes the S–H to return to sample (track) and acquire the next channel. Then after the acquisition is completed,

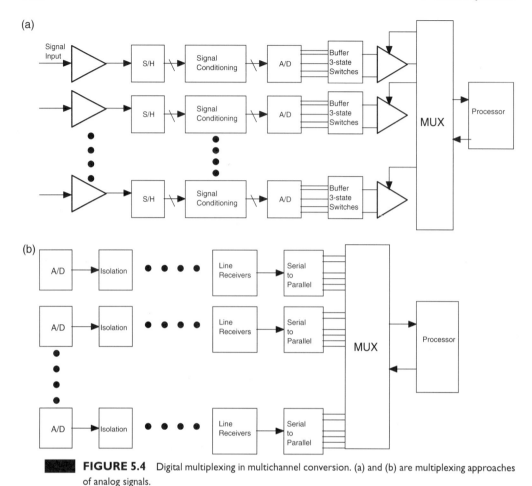

FIGURE 5.4 Digital multiplexing in multichannel conversion. (a) and (b) are multiplexing approaches of analog signals.

either immediately or on command, the sample–hold is switched to hold, a conversion begins, and the multiplex switch moves on.

5.4. HIGH-SPEED SAMPLING IN ADCs

Due to the various design constraints, it is really not possible to make the input of a high-speed ADC totally well behaved—with a high-input impedance, low capacitance, ground referenced, and free from transients. This means that the ADC drive amplifier must provide excellent ac performance. The tendency toward single-supply high-speed designs provides additional constraints. The input voltage range of high-speed single-supply ADCs may not be ground referenced; therefore, level shifting with single-supply op-amps is usually required.

Modern signal processing applications require ADCs that have wide dynamic range, high bandwidth, low distortion, and low noise. ADCs with

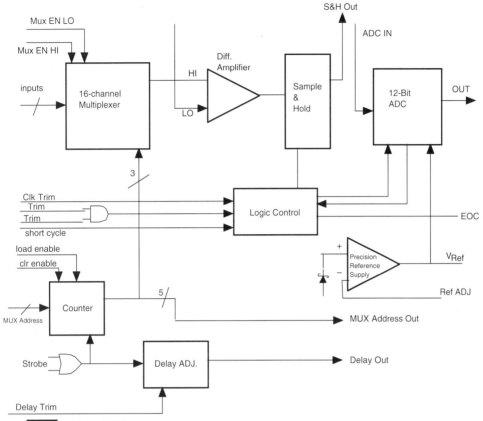

FIGURE 5.5 Early data acquisition system architecture.

internal sample–hold functions are generally specified in terms of familiar terms such as signal-to-noise ratio (SNR), signal-to-noise-plus-distortion (S/(N+D)), effective number of bits (ENOB), harmonic distortion, total harmonic distortion (THD), intermodulation distortion (IMD), and spurious free dynamic range.

In ADCs, an ideal sampling at a rate of F_s causes a quantization noise having an rms value of $q/\sqrt{12}$ measured in the Nyquist bandwidth dc to $fs/2$, where q is the weight of the LSB. To obtain q, we divide the full-scale input range of the ADC by the number of quantization levels, 2^N. For example, an ideal 10-bit ADC with a 2.048-V peak-to-peak input range has $2^{10} = 1024$ quantization levels, an LSB of 2 mV, and an rms quantization noise of $2\,\text{mV}\sqrt{12} = 577\,\mu V$.

Another method to express quantization noise is to convert it to a signal-to-noise ratio using the well-known expression:

$$S/N = 6.02N + 1.76\,\text{dB} \qquad (5.4)$$

An ADC produces noise as well as distortion products caused by a nonlinear transfer function. A fast Fourier transform (FFT) is used to calculate the rms value of all the distortions and noise products, and the actual signal-to-noise-plus-distortion, $S/(N+D)$, is computed. Solving Eq. (5.4) for N yields the expression for the effective number of bits (ENOB) as

$$\text{ENOB} = \frac{S/(N+D)_{\text{Actual}} - 1.76\,\text{dB}}{6.02} \quad (5.5)$$

Just about all ADCs have some sort of nonlinearities that contribute to non-ideal low-frequency performance, and this performance degrades as the input frequency increases. A good way to evaluate the performance of an ADC is to plot signal-to-noise-plus-distortion, $S/(N+D)$, as a function of input frequency.

If we get to plot the gain of an amplifier with a small signal of a few millivolts, we find that as the input frequency increases, there is a frequency at which the gain drops by 3 dB. This frequency is the upper limit of the small-signal bandwidth of the amplifier and it is dictated by the internal pole(s) in the amplifier response. If we drive the amplifier with a large signal so as to get full peak-to-peak output voltage, the upper 3-dB point is found at a lower frequency, limited only by the slew rate of the amplifier output stage. The high-level 3-dB point is known as the large-signal bandwidth of an amplifier. The large-signal bandwidth is an uncertain parameter in an amplifier because it depends on many uncontrolled variables, such as power supply, output amplitude, and the load. For situations where the large-signal bandwidth is less than the small-signal bandwidth, it is much better to define the output slew rate and calculate the maximum output swing of a given frequency. The large-signal bandwidth tells us the frequency at which the amplitude response of the ADC drops by 3 dB, but it does not tell us the relationship between distortion and frequency. It is easy to notice, however, that in general, noise and distortion increase with increasing frequency, and this reduces the resolution that we can obtain from the ADC.

A graph of the ratio of signal-to-noise-plus-distortion, $S/(N+D)$, with respect to input frequency is shown in Figure 5.6.

As seen in the figure, the SNR of a perfect N-bit ADC (with a full-scale sine wave input) is $6.02N + 1.76$ dB.

Today's ADCs use sampling, which means that the S–H architecture is part of the ADC. The aperture jitter of the S–H cannot be specified in specs sheets. The use of an additional high-performance S–H architecture can sometimes

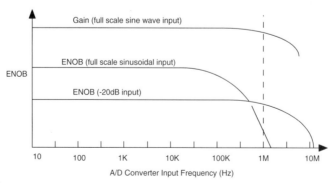

FIGURE 5.6 Signal-to-noise ratio vs frequency in amplifiers.

improve. The high-frequency ENOB of a sampling ADC can be more cost effective than replacing the ADC with a more expensive one.

There is also a fixed component that makes up an ADC aperture time. This component is called the effective aperture delay time and does not produce an error. It results in a time offset between the time the ADC is asked to sample and when the sampling actually takes place (Figure 5.7).

The distortion produced by an ADC cannot be analyzed in terms of second- and third-order intercepts as it is done with amplifiers. The reason is that there are two components of distortion in a high-performance data converter. One component is due to the nonlinearity associated with the analog circuits within the converter. This nonlinearity is shown in Figure 5.8.

The distortion associated with this type of nonlinearity is referred to as soft distortion and produces low-order distortion products. In a practical data converter, however, the soft distortion is usually much less than the other

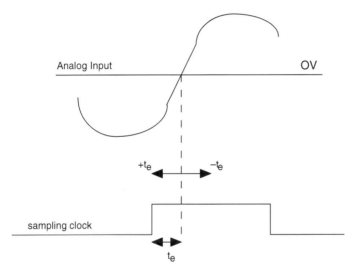

FIGURE 5.7 Aperture delay time in the sampling process.

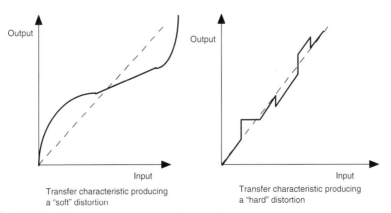

FIGURE 5.8 Nonlinearity associated with an ADC.

component of distortion, which is due to the differential nonlinearity of the transfer function and the cause of hard distortion, as shown in the figure.

5.5. SELECTION OF DRIVE AMPLIFIER FOR ADC PERFORMANCE

The ADC drive amplifier performs several functions. It isolates the signal source and provides a low-impedance drive to the ADC input. A low-impedance dc and ac drive source is important because the input impedance of the ADC may be signal dependent, and also the input may generate transient load currents during the conversion process. The drive amplifier provides the required gain and level shifting to match the signal to the A–D input voltage range.

The plot of $S/(N+D)$ for an ADC should be used as the main criterion for the drive amplifier. If the total harmonic distortion plus noise of the amplifier is 6–10 dB better than $S/(N+D)$ over the frequency range of interest, then the overall degradation in $S/(N+D)$ caused by the amplifier will be limited to between 0.5 and 1 dB, respectively. An example of this is shown in Figure 5.9 with the Analog Devices AD9022, which contains a three-pass subranging architecture and digital error correction.

The analog input is passed to the sampling bridge of the first internal S–H amplifier. The held value of the first S–H is applied to a 5-bit flash converter and a second S–H. The 5-bit flash converter resolves the most significant bits. These 5 bits are reconstructed via a 5-bit DAC and subtracted from the original S–H output signal to form a residue signal. A second S–H holds the amplified residue signal while it is encoded with a second 5-bit flash ADC. As before, the 5 bits are reconstructed and subtracted from the second S–H output to form a residue signal. This residue is amplified and encoded with a 4-bit flash ADC to provide the 3 LSB of the digital output and one bit of error correction. The digital error correction logic combines the data from the three flash converters and then presents the results as a 12-bit parallel digital word. The output stage is either TTL or ECL. Output data can be strobed on the rising edge of the encode command. In Figure 5.10 we observe the total harmonic distortion plus noise (THD+N) of the AD9631 drive amplifier superimposed on the $S/(N+D)$ plot

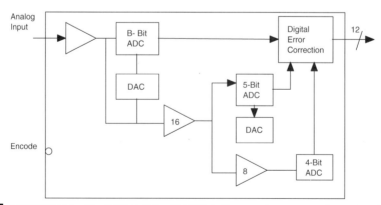

FIGURE 5.9 The AD9022 12-bit, 20-MSPS (million samples per second) sampling ADC.

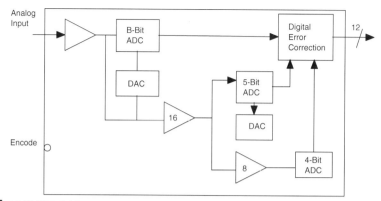

FIGURE 5.10 Total harmonic distortion plus noise effect on the AD9631.

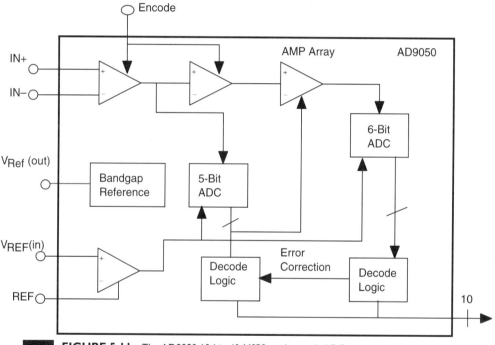

FIGURE 5.11 The AD9050 10-bit, 40-MSPS single supply ADC.

for the AD9022. Notice that the amplifier THD+N is at least 10 dB better than the ADC $S/(N + D)$ for input frequencies up to about 10 MHz (the Nyquist frequency).

Some ADCs require only a single supply. An example of such is the Analog Devices AD9050, 10-bit, 40-MSPS ADC designed for wide dynamic range such as ultrasound. A block diagram of the AD9050 is shown in Figure 5.11 and illustrates the two-step subranging architecture. The analog input circuit of the AD9050 is differential but it can be driven single-ended or differentially with equal performance. The input circuit of the AD9050 is a benign and constant 5 KΩ in parallel with approximately 5 pF. Because of its well-behaved input,

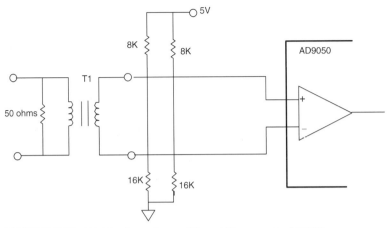

FIGURE 5.12 Wideband transformer differential inputs to the AD9050.

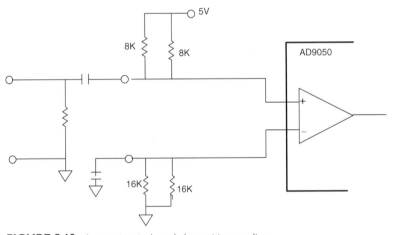

FIGURE 5.13 Input using single-ended capacitive coupling.

the AD 9050 can be driven directly from 50-, 75-, or 100-Ω sources without the need for a low-distortion buffer amplifier. In ultrasound applications it is normal to ac couple the signal (generally around 1 MHz) into the AD9050 differential inputs using the wideband transformer as shown in Figure 5.12.

If the input signal comes directly from a 50-, 75-, or 100-Ω single-ended source, capacitive coupling can be used, as shown in Figure 5.13.

5.6. DRIVING ADCs WITH SWITCHED CAPACITOR INPUTS

Many ADCs have switched capacitor input circuits. The effective input impedance can be a function of the sampling rate. The switches (usually CMOS) can inject charge on the ADC analog inputs. In this case, the internal sample end hold amplifier may generate a current spike on the analog input when the S–H circuit goes from the sample mode to the hold mode. Other spikes may be

DRIVING ADCs WITH SWITCHED CAPACITOR INPUTS

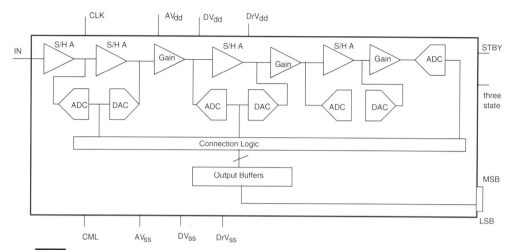

FIGURE 5.14 The AD876 ADC with switch capacitor input circuit.

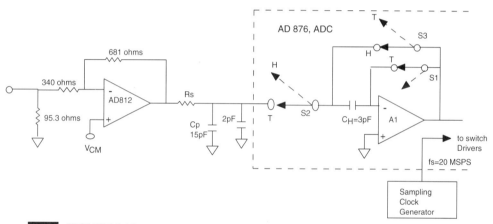

FIGURE 5.15 Implementation of the AD876 ADC.

generated during the actual A–D conversion. The current spikes on the output of the external ADC drive amplifier can produce voltage spikes, and conversion errors can result if the amplifier settling time is not adequate.

The Analog Devices AD876 ADC is a 10-bit, 20-MSPS low-power CMOS ADC with a switched capacitor sample and hold input circuit. An overall block diagram of the ADC is shown in Figure 5.14.

Operation of the AD876 switched capacitor input circuit is illustrated in Figure 5.15 and the associated waveforms in Figure 5.16. The CMOS switches $S1$, $S2$, and $S3$ control the action of the internal sample and hold.

Switches shown in Sample Mode
Switching Sequence.
$T \rightarrow H$: $S1$ opens, $S2$ opens, $S3$ closes.
Hold: conversion occurs.

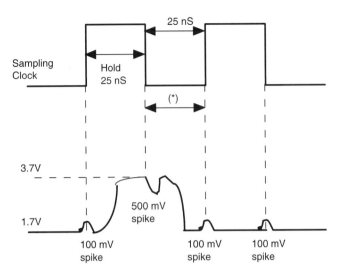

FIGURE 5.16 Associated waveform for the AD876 ADC.

$H \rightarrow T$: S3 opens, S1 closes, S2 closes.
C_H charged to new value.

5.7. GAIN SETTING AND LEVEL SHIFTING

For dc-coupled applications, the drive amplifier must provide the required gain and offset to match the signal to the input voltage range of the ADC. Figure 5.17 summarizes several gain and level shifting options.

A practical example of a single-supply video signal processing digitizing circuit is shown in Figure 5.18. The AD876 ADC operates on a single 5-V supply and its nominal input voltage range is 2 V peak-to-peak centered around an allowable common mode voltage between 2.7 and 3.1 V. The input voltage range of the AD876 is set by external references. The AD812-driven amplifier is a fast video op-amp with a common mode input voltage range of 1- to 4-V and 1- to 4-V output voltage range. In this amplifier–ADC combination, optimum performance is obtained by setting the AD876 input common mode voltage of 2.7 V, which corresponds to an upper and lower input range of 3.7 and 1.7 V, respectively.

The Thevenin equivalent circuit of the video signal is a ground-referenced, 0- to 2-V source with a 75-Ω source impedance to match the 75-Ω coaxial cable impedance to produce a standard 0- to 2-V video signal level of the load termination R_T. The AD812 functions as an inverter level shifter. The feedback resistor $R2$ is chosen to be 681 Ω for optimum flatness over the video bandwidth as recommended by the data sheet. The feed forward resistor, $R1$, is selected to give a signal gain of -2. The termination resistor R_T is chosen such that the parallel combination of R_T and R_1 is 75 Ω. The common mode voltage V_{cm} is determined by the voltage divider formed by $R2$, $R1$, and RT and the 75-Ω

GAIN SETTING AND LEVEL SHIFTING   121

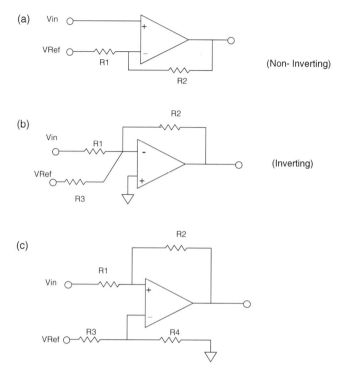

FIGURE 5.17 (a) through (c) shows several level shifting options in amplifiers.

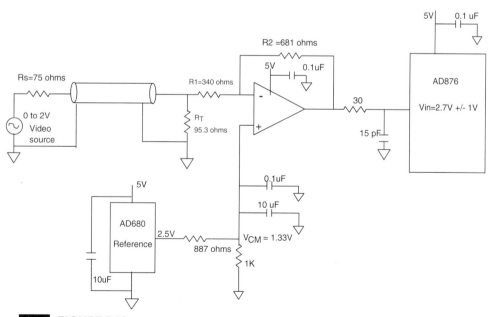

FIGURE 5.18 Example of simple video processing.

resistance and given by the expression

$$V_{cm} = 3.7 \left\{ \frac{R_s \| R_T + R1}{R_s \| R_T + R1 + R2} \right\} = 3.7 \left\{ \frac{42 + 340}{42 + 340 + 681} \right\} = 1.33\,\text{V} \quad (5.6)$$

where the 3.7 V is the corresponding op-amp output voltage when the video signal is 0 V. The 1.33-V common mode voltage is derived from the AD680 22.5-V reference using a resistor divider and is decoupled with a 10-μF capacitor in parallel with a 0.1-MF low-inductance ceramic capacitor.

5.8. HIGH-SPEED SAMPLING ADC EXTERNAL REFERENCE VOLTAGE GENERATION

Due to several constraints, it is not possible to integrate the reference and the ADC on the same IC, and an external reference voltage is needed. There are several types of reference ICs such as 1.25, 2.5, 3, 3.3, and 4.5 V. The AD876 requires two references: one for each end of its input range, which is nominally set for 1.7 and 3.7 V. The output impedances of the drive sources must be low at high frequencies to absorb the transient current generated by the ADC reference input terminals.

The circuit configuration shown in Figure 5.19 uses the REF198 (4.096 V) reference and a dual FET-input single supply op-amp (AD 822) to generate the two voltages. The AD876 reference inputs each have a force (F) and

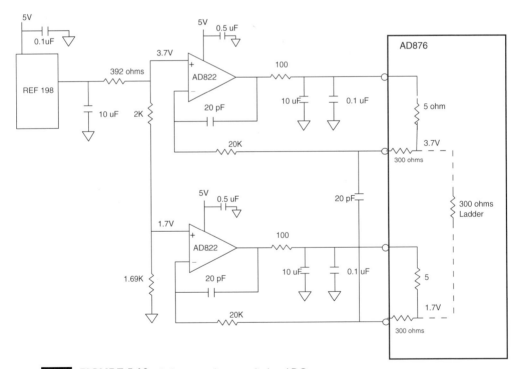

FIGURE 5.19 Reference voltages applied to ADCs.

ADC INPUT PROTECTION

(SENSE) (S) pin. The Kelvin connection compensates for the voltage drop in the internal parasitic resistances. The internal ADC reference ladder impedance is approximately 300 Ω, which requires the AD822 to source and sink about 6.7 MA. The two reference force pins are decoupled at low and high frequencies. The 20-μF capacitor across the two sense pins adds additional decoupling for differential transients. The AD822 must be compensated to drive the large capacitive load.

5.9. ADC INPUT PROTECTION

There is a need to protect the input to a high-speed ADC from overdrive. The analog input should never exceed the supply voltage by more than 0.3 V. In a dual-supply system, this rule applies to both supplies. In some ADCs, the analog input is protected internally.

In these cases, an external resistor is required to limit the input current to 5 mA or less under the overvoltage condition. Several overdrive protection schemes using external diodes are shown in Figure 5.20.

Instead of diodes, a clamping amplifier can be used, as shown in the circuit of Figure 5.21. In the AD8037 clamping amplifier, the clamping voltages are set to 0.55 and −0.55 V, referenced to the ±0.5-V input signal, with the external resistive dividers. The AD8037 also supplies a gain of 2, and an offset of −1 V (using the AD780 voltage reference), to match the 0- to −2-V input range of the AD9002 flash converter. The output signal is clamped at 0.1 and −2.1 V. This multifunction clamping circuit therefore performs several important functions as well as preventing damage to the flash converter, which occurs when its input exceeds 0.5 V, thereby biasing the substrate diode.

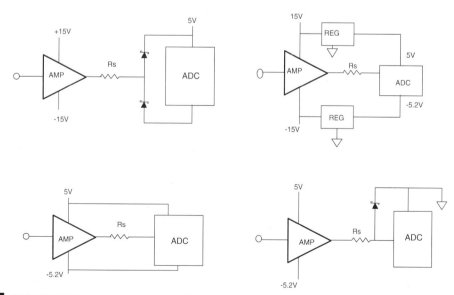

FIGURE 5.20 Overdrive protection for ADCs.

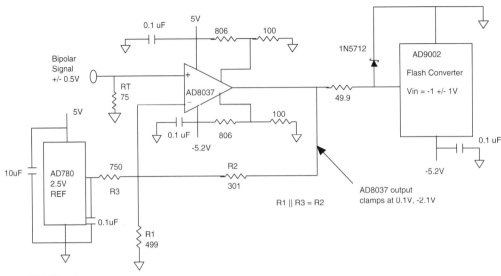

FIGURE 5.21 Clamping amplifier for an ADC.

The feedback resistor is chosen for optical bandwidth to obtain a gain of 2.

$$\frac{R_1 R_3}{R_1 + R_3} = R_2 = 301\,\Omega$$

In addition, the Thevenin equivalent output voltage of the AD780 is a 2.5-V reference and the R_3–R_1 divider must be 1 V to provide −1-V offset at the output of the AD8037:

$$\frac{(2.5)R_1}{R_1 + R_3} = 1\,\text{V}$$

Solving the two simultaneous equations yields $R_1 = 499$ and $R_3 = 750\,\Omega$.

5.10. NOISE CONSIDERATIONS IN HIGH-SPEED SAMPLING ADCs

Due to the wide bandwidth front ends of ADCs, the high-speed sampling of an ADC is susceptible to coupled noise. There are many sources for such noise (see Figure 5.22). In high-speed systems, where the resistors of the source and the op-amp feedback network rarely exceed 1 KΩ, the resistor Johnson noise can often be neglected. In the case of voltage feedback op-amps the input current noise can often be neglected. For current feedback op-amps, the inverting input current noise generally dominates. At higher noise gains, the effects of voltage noise become significant.

NOISE CONSIDERATIONS IN HIGH-SPEED SAMPLING ADCs

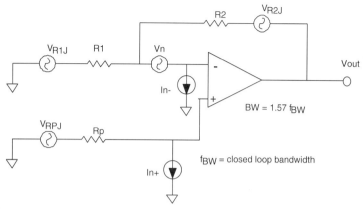

FIGURE 5.22 Noise in operational amplifiers.

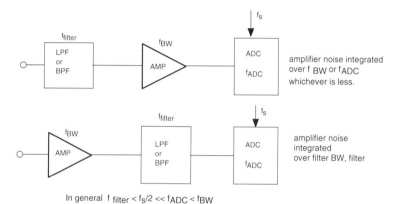

FIGURE 5.23 Filtering noise out of op-amps.

$$V_{out} = \sqrt{BW} \left[I_{n-}^2 R_2^2 + I_{n+}^2 R_p^2 \left(1 + \frac{R_2}{R_1}\right)^2 + V_n^2 \left(1 + \frac{R_2}{R_1}\right)^2 \right.$$
$$\left. + 4KTR_2 + 4KTR_1 \left[\frac{R_2}{R_1}\right]^2 + 4KTR_p \left(1 + \frac{R_2}{R_1}\right)^2 \right]^{1/2} \quad (5.7)$$

The bandwidth for integration is either the op-amp closed-loop bandwidth, or the ADC input bandwidth. In most cases, the input of the ADC acts as a low-pass filter to the op-amp noise. The proper positioning of an antialiasing filter can reduce the effects of the op-amp noise, as shown in Figure 5.23.

Another method for noise control is the use of proper supply decoupling techniques. An example of this technique is shown in Figure 5.24. The power supply input is first decoupled to the large-area low-impedance ground plane with a good quality, low ESL and low ESR tantalum electrolytic capacitor. This capacitor bypasses low-frequency noise to the ground plane. The ferrite bead reduces high-frequency noise to the rest of the circuit. A low-inductance

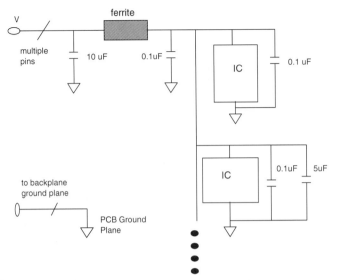

FIGURE 5.24 Bypass capacitance for IC chips in PCBs.

ceramic capacitor at each power pin on each IC should be used. Surface-mount chip capacitors for minimum inductance should also be used. Some ICs may require an additional small tantalum electrolytic capacitor.

It is recommended that multilayer boards be used with one layer totally dedicated to ground at least. When connecting to the back plane, use around 30 to 40% of the pins on each printed circuit board (PCB) connector for the ground to maintain a low-impedance ground plane between the various PCBs in a multiboard system.

It should be a good practice to physically separate analog-sensitive components from digital components. It is usually a good practice to establish separate analog and digital ground planes or each PBC, as shown in Figure 5.25.

The separate analog and digital ground planes are continued on the backplane using either motherboard ground planes or wire screens. The ground planes are joined together at the system chassis ground or single-point ground located at the common return point for the power supply. The diodes are inserted to avoid accidental dc voltages from developing between the two ground systems. The ADCs (even those with mixed-signal ICs) should be addressed as analog circuits and also grounded and decoupled to the analog ground plane. Figure 5.26 shows the proper grounding of ADCs and DACs.

There is nothing at this point that the designer can do to avoid the effects of wire-bond inductance and resistance associated with connecting the pads on the chip to the package pins. There will be some coupling through the stray capacitance C_{stray} between the digital and analog sides. In addition there is some stray capacitance between pins in the IC package.

The logic supply pin V_{cc} can be isolated by the insertion of a small ferrite bead, as shown in Figure 5.27. The internal digital currents of the ADC will return to ground through the V_D pin decoupling capacitor and will not appear in the external ground circuit. It is also desirable to place a buffer latch adjacent

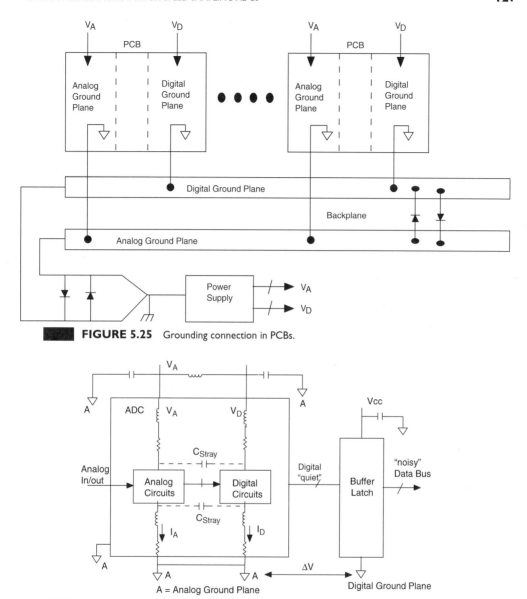

FIGURE 5.25 Grounding connection in PCBs.

FIGURE 5.26 Proper grounding of ADCs and DACs.

to the converter to isolate the converter's digital lines from any noise that may be found on the data bus. The buffer latch and other digital circuits should be grounded and decoupled to the digital ground plane of the PCB. Any noise between the analog and digital ground reduces the noise margin at the ADC digital interface. Since the digital noise immunity is of the order of thousands of millivolts, it does not provide concern.

The sampling clock generator (OSC) should also be grounded and heavily decoupled to the analog ground plane. A low-noise crystal oscillator should be used to generate the clock. Sampling clock jitter modulates the input signal and

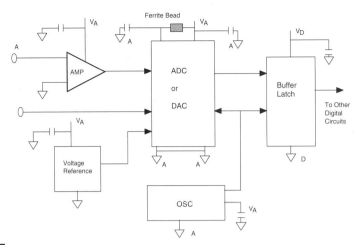

FIGURE 5.27 Grounding and ferrite usage for noise control in IC components.

raises the noise and distortion floor. The sampling clock generator should be isolated from noise digital circuits and grounded and decoupled to the analog ground plane. Separated power supplies for analog and digital circuits are also highly desirable but not always affordable. All converter power pins should be decoupled to the analog ground plane and all logic circuit power pins should be decoupled to the digital ground plane. When a ground plane is used, it should act as a shield for sensitive signals. Figure 5.28 shows a good layout for a data acquisition system where all sensitive areas are isolated from each other and signal paths are kept as short as possible.

5.11. MULTICHANNEL APPLICATIONS FOR DATA ACQUISITION SYSTEMS

Data acquisition systems involve digitizing analog signals using analog-to-digital converters. The ADCs are followed by a digital processor that performs the needed data analysis. In a process control application, the process controller generates feedback signals that must be converted back into analog form using a DAC. The term *data acquisition* refers to a multichannel system. Through the feedback from the digital processor, DACs convert the digital response into analog. A data acquisition system is shown in Figure 5.29 in which each channel has its own dedicated ADC and DAC.

As shown in the figure, multiplexers are a fundamental block of the data acquisition system. A simplified diagram of an analog multiplexer is shown in Figure 5.30. The number of input channels ranges from 4 to 16. Some multiplexers have internal channels-address-decoding logic and registers, while others lack these functionalities and have to be implemented externally. Unused multiplexer input must be grounded. Multiplexer switching times range from about 50 ns to over 1 μs, the on resistance (R_{on}) can range from 25 Ω to several hundreds ohms, and off-channel isolation from 50 to 90 dB. The multiplexer output should be isolated from the load by the usage of a buffer amplifier. An

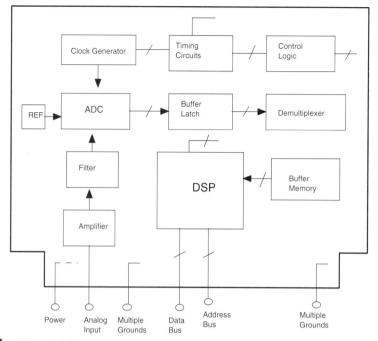

FIGURE 5.28 A PCB layout for good signal routing.

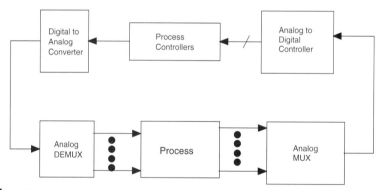

FIGURE 5.29 Simplified data acquisition system.

m-channel multiplexed data acquisition system is shown in Figure 5.31. The multiplexer output drives a programmable buffer amplifier whose gain can be adjusted, depending on the channel signal level. The maximum sampling frequency and the maximum input frequency are limited by the multiplexer switching time, the buffer amplifier settling time, and the ADC conversion time as shown by the formulas

$$f_s \leq \frac{1}{t_{\text{conv}} + \sqrt{t_{\text{MUX}}^2 + t_{\text{amp}}^2}} \qquad f_{\text{in}} \leq \frac{1}{\pi 2^N T_{\text{conv}}}$$

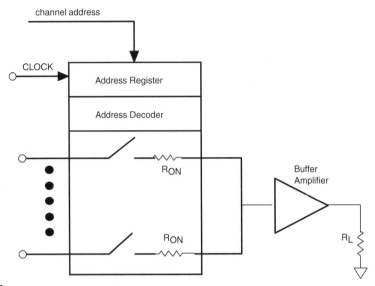

FIGURE 5.30 Illustration of a multiplexer.

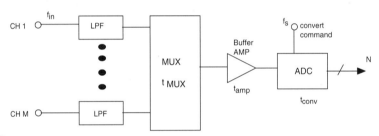

FIGURE 5.31 Multiplexed data acquisition system.

Adding sample–hold to the ADC, as shown in Figure 5.32, allows the processing of much faster signals with almost no increase in system complexity. Filtering in the data acquisition system prevents aliasing of unwanted signals, but the noise is also reduced by limiting the bandwidth. The filtering at the input of each channel is used to prevent aliasing of signals that are outside the Nyquist bandwidth. The channel sampling rate is given by f_s/M. In general

$$\sqrt{t_{\text{MUX}}^2 + t_{\text{amp}}^2} \ll t_{\text{acq}} + t_{\text{conv}}$$

Therefore,

$$f_s \leq \frac{1}{t_{\text{acq}} + t_{\text{conv}}}$$

And the corresponding Nyquist frequency is $f_s/2M$. The filter should provide sufficient attenuation of $f_s/2M$.

There are certain applications where it is required to sample a number of channels simultaneously. A typical configuration is shown in Figure 5.33. Each channel requires its own filter and S–H. Each sample–hold is simultaneously

EXTERNAL PROTECTION OF AMPLIFIERS

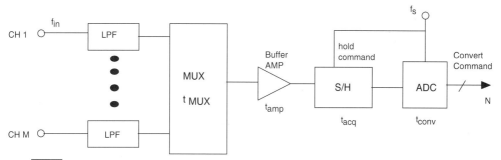

FIGURE 5.32 Multiplexed data acquisition with a sample–hold circuitry system.

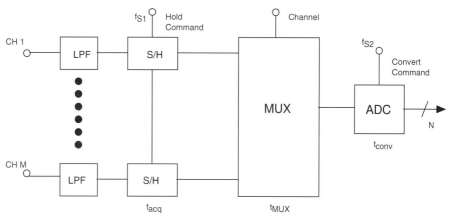

FIGURE 5.33 Sampling multiple channels.

placed in the hold mode by a common command signal. During the input S–H hold time, the multiplexer is sequentially switched from channel to channel, and the single nonsampling ADC is used to digitize the signal on each channel.

If a sampling ADC is used to perform the conversion the acquisition time of the second S–H (see Figure 5.34) t_{acq2} should also be considered in determining the maximum ADC sampling rate f_{s2}:

$$f_{s2} \leq \frac{1}{t_{acq2} + t_{conv}} \qquad f_{s1} \leq \frac{1}{t_{acq1} + M/F_{s2}}$$

5.12. EXTERNAL PROTECTION OF AMPLIFIERS

Often amplifiers are used in applications for data acquisition and processing where their inputs are exposed to electromagnetic interference, electrostatic discharge, and overvoltage events.

If an amplifier's input voltage is higher than the supply voltage, the unit amplifier can be damaged, usually by high current flow, even when turned off. The typical maximum input voltage allowed for both the positive and negative supplies is about 0.3 V outside the supply voltage. To avoid damage to the

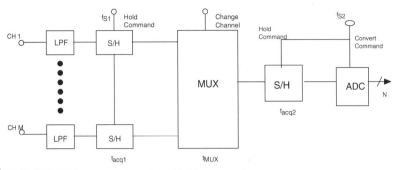

FIGURE 5.34 A second sample and hold circuit in data acquisition.

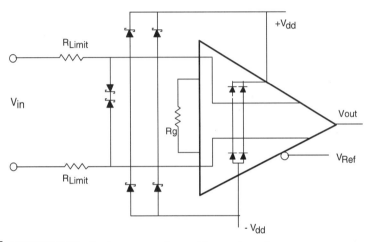

FIGURE 5.35 Diode protection for amplifier supplier voltage.

amplifier input, clamping to limit the input current to either 10 or 20 mA is used. This value is a conservative rule of thumb based on metal trace widths in a typical amplifier input stage. Higher currents can cause metal migration, which will eventually lead to an open trace. Therefore, limiting the current is very important to guarantee long-term reliability.

Figure 5.35 shows an equivalent input circuit for the input stage of an amplifier. The amplifier has internal resistance in series with input transistor junctions and their protective diodes. The internal diodes protect the unit from input voltages up to 10 voltages greater than supply voltage. So for ±15 V, the maximum safe level is around 25 V.

In addition, the differential input voltage should also be given a value that limits the maximum input current to 10 mA. The circuit in Figure 5.36 shows that the input current flows through two external R_{limit} resistors, the two internal R_s resistors, the gain setting R_g, and two diode drops (D1 and the V_{be} junction of $Q2$).

For a given differential input voltage, the input current is a function of R_g. A generalized external current limiting resistor can ensure input protection (Figure 5.36). The amplifier in Figure 5.36 is protected against both differential and common-mode voltages. If the amplifier has internal protection diodes

EXTERNAL PROTECTION OF AMPLIFIERS   133

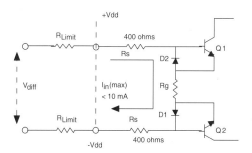

$$V_{diff} = I_{in}(2R_S + 2R_{Limit} + R_g) + 1.2V$$

$$V_{diff(max)} <= I_{in(max)}(2R_S + 2R_{Limit} + R_g) + 1.2V$$

FIGURE 5.36 Limiting maximum input current.

to the supplies (as shown in Figure 5.36), the diodes conduct at about 0.6-V forward drop above or below the supply rails. Therefore, the internal diodes have dual roles: protect against electrostatic discharge (ESD) and clamp the input voltage to 0.6 V beyond the supply rails.

Limiting the external current so that the maximum input current is not more than 20 mA may require large values of R_{limit} and this can also result in increased resistor (Johnson) noise. For example, resistors contribute noise according to the equation

$$\text{noise}(nV/\sqrt{Hz}) = \sqrt{4KRT} \times 10^9 \tag{5.8}$$

where K is Boltzmann's constant (1.38×10^{-23} J), R is the resistance in ohms, and T is the temperature in Kelvin (~ 300 at room temperature). For example, a 1 KΩ resistor has a Johnson noise of $4nV/\sqrt{Hz}$ at room temperature. Since the protection circuit includes two equal resistors, whose noise is uncorrelated, which means the two noise sources are independent of each other, the preceding results must be multiplied by the square root of 2 (the root sum square of the two noise voltages). A reasonable balance between the protection provided and the increased resistor noise is introduced.

For cases where the required protection resistor generates too much noise, external Schottky protection diodes, as shown in Figure 5.36, are used. The diodes begin to conduct at about 0.3 V so the overvoltage current is shunted through them to the supply rails rather than through the internal diodes. Therefore, you can set R_{limit} by the maximum allowable diode current, which can be much larger than the internal limit of 10 or 20 mA.

Electrostatic Discharge: The high voltages and high peak currents generated by ESD can partially or permanently damage an IC. Figure 5.37 shows a simple technique for protecting amplifiers against high-voltage ESD. Carbon resistors, which are non-inductive, should be used as protection resistors instead of devices made of metal film or carbon film.

Electromagnetic Interference: Radio frequency interference can seriously affect the dc performance of high-accuracy circuits. Because of their low bandwidth, amplifiers do not accurately amplify signals in the megahertz range.

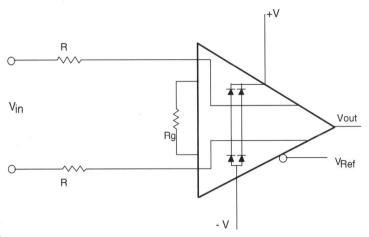

FIGURE 5.37 Protection against ESD in amplifiers.

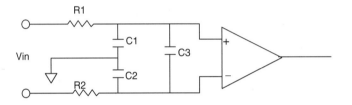

FIGURE 5.38 Filtering out noise in amplifiers.

These out-of-band signals (either differential mode or common mode) can couple into the precision amplifier through its input, output, or power-supply pins. Then, through various junctions in the amplifier, an unwanted dc offset can appear. Many types of precision amplifiers are specially susceptible to common mode RFI. Filtering considerations as shown in Figure 5.38 should be appropriate. Common mode filtering ($R1/C1$, $R2/C2$) represented by τ_{diff} and differential mode filtering ($R1 + R2$, $C3$) represented by τ_{cm} are shown, where $R1 = R2$, $C1 = C2$, $\tau_{\text{diff}} = (R1 + R2)C3$, $\tau_{cm} = R1 * C1 = R2 * C2$, $\tau_{\text{diff}} \gg \tau_{cm} R1 * C1$ should match $R2 * C2$.

$$\text{differential filter} = \frac{1}{2\pi(R1 + R2)[(C1C2/C1 + C2) + C3]}$$

Common mode chokes offer an alternative to RC filters. Their dc resistance is low (a few ohms). They attenuate RFI and add very little noise as compared to RC networks. The selection of the common mode choke is very important. Figure 5.39 shows the implementation of such a choke. The figure also shows an RC filter to protect the amplifier output from RFI. A ferrite band in series with the output is the simplest output filter.

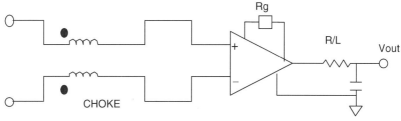

FIGURE 5.39 Choke to remove common mode noise.

5.13. HIGH-SPEED ADC ARCHITECTURES

There are basically four types of high-speed converter architectures addressed in this approach: (1) flash, (2) successive approximations, (3) subranging, and (4) digitally corrected subranging.

5.13.1. Basic Flash Converter Operation

Recent advances have characterized the flash converters as having high sampling rates and the ability to convert fast video input signals, usually without requiring a separate sample–hold amplifier (limited capability). A block diagram of a typical flash converter is shown in Figure 5.40. The analog input signal to be digitized is sent simultaneously to $2^N - 1$ latched comparators, where N is the number of bits. The timing diagram of some of the signaling is shown in Figure 5.41.

Flash converter "full-power bandwidth" (FPBW) typically indicates the maximum frequency at which the amplifier is capable of producing the maximum specified peak-to-peak output voltage at some level of distortion. Another commonly used definition is to calculate FPBW from the slew rate (SR) of the amplifier using the equation

$$\text{FPBW} = \frac{\text{SR}}{2\pi V_0} \tag{5.9}$$

where $\pm V_0$ is the output voltage range of the amplifier.

The reference voltage for each comparator is one least significant bit higher than the comparator immediately below it. When an analog signal is present at the input of the comparator bank, all the comparators whose reference voltage is below the level of the input signal will assume a logic "L" output. The comparators that have their reference voltage above the input signal will assume a logic "0" output. The decoding results in a binary digital output.

5.13.2. Driving Flash Converters

The flash converter usually comes from 50-, 75-, or 93-Ω sources. The signal may be bipolar or unipolar. If the input range of the flash converter is not compatible with the signal, a wideband amplifier will be required to produce the required gain and offset.

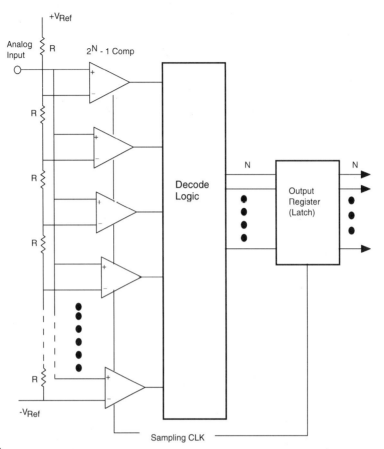

FIGURE 5.40 Block diagram of a flash converter.

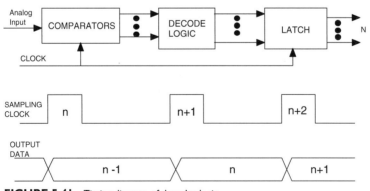

FIGURE 5.41 Timing diagram of decoder logic.

The reference voltage for each comparator is one least significant bit higher than the comparator immediately below it. When an analog signal is present at the input of the comparator bank, all the comparators whose reference voltages are below the level of the input signal will assume a logic "1" output. The comparators that have their reference voltage above the input signal will assume a logic "0" output. The decoding results in a binary digital output (Figure 5.42).

The comparator bank in a flash converter has two states. In the first state, which is controlled by the sampling clock, the comparator tracks the analog input signal. In this state, the comparator outputs are changing and the binary decoding logic output is invalid. When the sampling clock changes to the opposite logic level, the comparators are latched or "held," much the same as in a sample–hold amplifier.

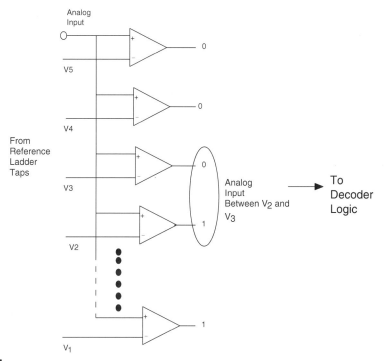

FIGURE 5.42 Illustrating comparator output.

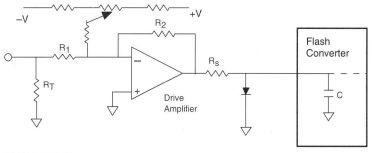

FIGURE 5.43 Driving flash converters.

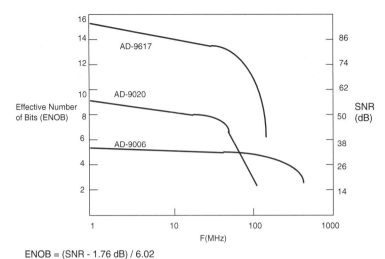

ENOB = (SNR - 1.76 dB) / 6.02

FIGURE 5.44 Harmonic distortion of a typical amplifier.

For some flash converters, the input capacitance is so high that a buffer amplifier is required to preserve the signal bandwidth, as shown in Figure 5.43. The value of the R_s resistor is dependent on the capabilities of the converters themselves. Since most applications do require a buffer ahead of the flash converter, the user must select it carefully. The primary consideration is to match the dynamic performance (harmonics, SNR, etc.) of the amplifier to that of the flash converter so that the performance of the flash converter is not degraded by the amplifier. Figure 5.44 shows the harmonic distortion of a typical amplifier.

5.13.3. Successive Approximation ADCs

This ADC architecture has been widely used in the industry. A block diagram of a successive approximation ADC is shown in Figure 5.45. The building blocks are made up of a capacitor, a DAC, and control logic (successive approximation register, or SAR). The overall static accuracy is primarily determined by the DAC.

The analog input drives the input of a high-precision comparator, and the DAC output is connected to the other input. The conversion technique consists of comparing the unknown input against a very defined voltage or current generated by the DAC. The input of the DAC is the digital number at the ADC output.

After the conversion command is applied and the converter is cleared the DAC most significant bit (MSB) output ($\frac{1}{2}$ full scale) is compared with the input. If the input is greater than the MSB, it remains on (i.e., "1" in the output register). If the input is less than the MSB it is turned off (i.e., "0" in the output register), and the next bit is tried. If the second bit does not add enough weight to exceed the input, it is left on ("1") and the third bit is tried. If the second bit tips the scale too far, it is turned off ("0") and the third bit is tried. The process continues in order of descending bit weight until the last bit has been

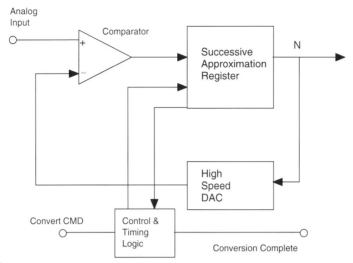

FIGURE 5.45 DAC architecture.

tried. Once the process is completed the conversion complete line changes state to indicate there was a valid conversion. The contents of the output register form a binary digital code corresponding to the input signal magnitude.

5.13.4. Subranging ADCs

A block diagram of a 12-bit subranging ADC is shown in Figure 5.46. It is composed of a 5-bit and an 8-bit flash converter. If there were no errors, the 5-bit "residue" signal applied to the 8-bit flash converter by the summing amplifier would never exceed one-half of the range of the 8-bit flash. The extra range in the second flash converter is used in conjunction with the error correction logic (usually just an adder) to correct the output data for most of the errors inherent in the traditional uncorrected subranging converter previously discussed.

When the analog signal being digitized by an ADC exceeds one-half of the sampling rate, the condition is called the "super-Nyquist" rate or "undersampling." The Nyquist criteria says that the bandwidth (not the frequency) of the signal being digitized should not exceed one-half the sampling rate for all information to be preserved.

As an example, consider the frequency division multiplexer data in Figure 5.47 occupying the bandwidth between 70 and 110 kHz which is sampled at a frequency of 112 kHz. The figure shows the spectrum of the signal and the location of the "aliasing" component. The receiver contains a bandpass filter which would be responsible for filtering out the aliasing terms. Operation of an ADC in the undersampling mode requires that the dynamic performance of the converter be known for input frequencies above $f_s/2$. The SNR, ENOB, and the harmonic performance of the converter degrades as the input frequency is increased and may cause the converter to be useless for undersampling applications.

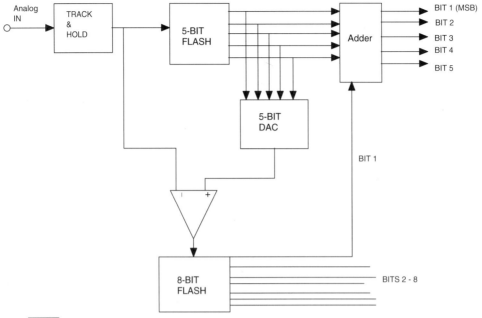

FIGURE 5.46 Subranging ADCs.

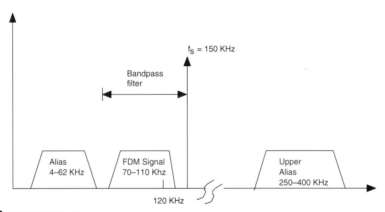

FIGURE 5.47 Sampling an analog signal without aliasing.

The technique of sampling a signal at greater than twice its maximum frequency is called oversampling. Increasing the sampling rate beyond the Nyquist rate of $f_s/2$ makes the design of the antialiasing filter much easier. This is shown in Figure 5.48.

The effective SNR can also be improved by oversampling for a given sampling rate f_s. The theoretical rms quantization noise in the bandwidth $f_s/2$ is given by $q/\sqrt{12}$, where q is the weight of the least significant bit. The theoretical formula for the SNR of an N-bit ADC is given by the expression

$$\text{SNR} = 6.02N + 1.76\,\text{dB} + 10\log(f_s/2f_a) \qquad (5.10)$$

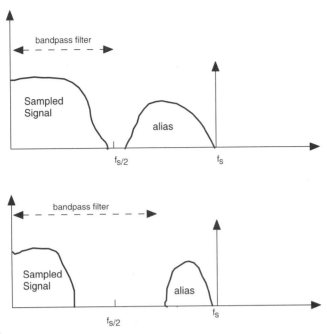

FIGURE 5.48 The oversampling technique.

where f_a is the analog bandwidth of interest. To obtain this improvement digital postprocessing of the quantized data in conjunction with a digital low-pass filter of bandwidth for f_a is implemented. The third term in the equation represents the increased SNR due to oversampling.

Dynamic or static errors are usually present in ADC. These dynamic errors increase as input slew rates becomes greater. The actual SNR measurement will, therefore, be less than theoretical. The ENOB is given by the equation

$$\text{ENOB} = \frac{\text{SNR}_{\text{actual}} - 1.76\,\text{dB} - 10\log(f_s/2f_a)}{6.02} \tag{5.11}$$

The noise is calculated using DFT techniques. It includes not only quantization noise but also the harmonics of the input sine wave.

REFERENCES

Burton, Phil. 1984. *CMOS DAC Application Guide*, 3rd ed. Analog Devices.
Ferguson, P. Jr., A. Ganesan, and R. Adams *et al.* 1991. "An 18-bit 20 KHz dual sigma delta A/D converter." *ISSCC Digest of Technical Papers*, February.
Harris, Steven. 1989. "The effects of sampling clock jitter on Nyquist sampling analog-to-digital converters and on oversampling delta sigma ADCs." Audio Engineering Society Reprint 2844, October.
High Speed Design Seminar. 1990. Analog Devices.

Koch, R., B. Heise, F. Eckbauer, E. Engelheardt, J. Fisher, and F. Parzefall. 1986. "A 12-bit sigma-delta analog-to-digital converter with a 15 MHz clock rate." *IEEE J. Solid State Circ.* SC-21 (6).

Y. Matsuya *et al.*, 1989. "17-bit oversampling D/A conversion technology using multistage noise shaping." *IEEE J. Solid State Circ.* 24(4), 969–975.

Sheingold, Daniel H., Ed. 1986. *Analog Digital Conversion Handbook*. Prentice-Hall, Englewood Cliffs, NJ.

6
NOISE AND INTERFERENCE ISSUES IN ANALOG CIRCUITS

6.1. BASIC NOISE CALCULATION IN OP-AMPs

The noise in operational amplifiers (op-amps) is related to the passive and active components within the circuit. It is also the kind of noise that could induce errors that could not be detected by dc error analysis. Noise can be random and repetitive, either a voltage or current form and can be at any frequency. Noise can be qualitatively classified as either white noise or color noise. Example of white noise are Johnson (or thermal) noise and shot noise that can exist up to a frequency of 100 GHz. Color noise has an amplitude that changes over frequency such as flicker noise $1/f$ or popcorn noise. An example of noise density spectrum is shown in Figure 6.1.

The noise spectral density is the rms value of the noise voltage V_n or a noise current I_n, which is expressed as a voltage or current per $\sqrt{\text{Hz}}$. The power spectral density is defined as the derivative of noise power over frequency range.

$$P(\text{W/Hz}) = \frac{dP_n}{df} \qquad (6.1)$$

The power spectral density for the voltage and current is defined as

$$V_n = \frac{V_n(\text{rms})}{\sqrt{\Delta f}} \ \text{V}/\sqrt{\text{Hz}}$$
$$I_n = \frac{I_n(\text{rms})}{\sqrt{\Delta f}} \ \text{A}/\sqrt{\text{Hz}} \qquad (6.2)$$

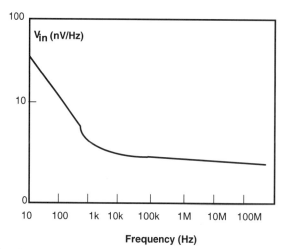
FIGURE 6.1 Example of noise density spectrum.

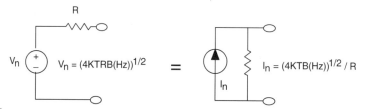
FIGURE 6.2 Thermal noise representation of a resistor.

6.1.1. Thermal Noise

In all electronic devices, thermal noise results from the random motion of free electrons in a conductor as a result of thermal agitation. Therefore, the thermal noise power is directly proportional to temperature and frequency

$$P_n = KTB(\text{Hz}) \text{ J/s} \tag{6.3}$$

where $K = 1.3805 \times 10^{-23}$ J/K is the Boltzmann's constant, T is the absolute temperature in Kelvin, and B (in hertz) is the bandwidth of the system. In conductors and semiconductors, the thermal noise is always present. For example, an ohmic resistor (see Figure 6.2) can experience a thermal noise voltage given by

$$V_n(\text{rms}) = \sqrt{4KTRB \text{ (Hz)}} \tag{6.4}$$

or in terms of spectral noise density,

$$\frac{V_n}{\sqrt{\text{Hz}}} = 4KTR \quad \text{nV}/\sqrt{\text{Hz}} \tag{6.5}$$

The noise figure in op-amps not only reflects the noise contributions of the IC itself but it also describes the IC with its feedback network, source, and

BASIC NOISE CALCULATION IN OP-AMPs

load resistance. With the use of noise figure a gain block can be completely characterized and total system noise calculations can be obtained by summing the noise figure of each stage.

The noise figure for an op-amp is the logarithm of the signal-to-noise ratio on the input of the amplifier to the signal-to-noise ratio at the output:

$$\text{noise figure} = \text{NF} = 10 \log \frac{(\text{SNR})_{\text{in}}}{(\text{SNR})_{\text{out}}} \tag{6.6}$$

In order to calculate the noise figure for op-amp gain stage the equation used is

$$\text{NF} = 10 \log \left(1 + \frac{V_n^2 + (I_n R_s)^2}{4KTR_s}\right) \tag{6.7}$$

It can be shown that the noise figure includes the voltage and current noise from the amplifier. The noise current inflows through the source impedance R_s. An important factor is the bandwidth. To calculate the total noise, the total output noise spectral density, which is given by nV/$\sqrt{\text{Hz}}$ is multiplied by the square root of the bandwidth. The calculation of op-amp noise in a size noninverting configuration is shown in Figure 6.3.

To obtain the total output noise, each term is multiplied by its gain and taken to the output as a voltage. Finally, all the terms are squared and added together and then take the square root of the sum of the squares. The individual

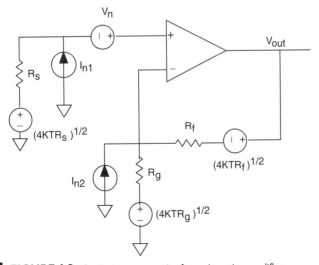

FIGURE 6.3 Intrinsic op-amp noise for an inverting amplifier.

terms are

$$R_s \rightarrow\rightarrow \sqrt{4KTR_s}\left(1 + \frac{R_f}{R_g}\right)$$

$$I_{n1} \rightarrow\rightarrow I_{n1}R_s\left(1 + \frac{R_f}{R_g}\right) \quad (6.8)$$

$$V_n \rightarrow\rightarrow V_n\left(1 + \frac{R_f}{R_g}\right)$$

$$I_{n2} \rightarrow\rightarrow I_{n2}R_f$$

$$R_g \rightarrow\rightarrow \sqrt{4KTR_g}\frac{R_f}{R_g} \quad (6.9)$$

$$R_f \rightarrow\rightarrow \sqrt{4KTR_f}$$

$$V_{out} = \left[(4KTR_s + (I_{n2}R_s)^2 + V_n^2)\left(1 + \frac{R_f}{R_g}\right)^2 + (I_{n1}R_f)^2 \right.$$
$$\left. + 4KTR_f\left(1 + \frac{R_f}{R_g}\right)\right]^{1/2} \quad (6.10)$$

6.2. FUNDAMENTAL OP-AMP SPECIFICATIONS

A block diagram of a basic op-amp is shown in Figure 6.4. The input stage is basically a differential input. Op-amps with a differential input as well as a differential output have very good common mode rejection ratio. The op-amp contains a high-gain stage with a single pole frequency response. The output is a single-ended output stage.

In the figure, $A(s)$ is known as the open-loop voltage gain and it is the gain with respect to the differential input voltage $V = (V_+ - V_-)$; $A(s)$ is a

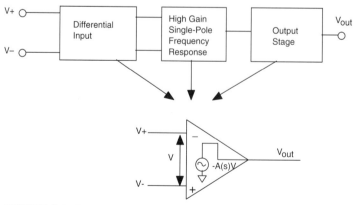

FIGURE 6.4 Basic block diagram of an op-amp.

FUNDAMENTAL OP-AMP SPECIFICATIONS

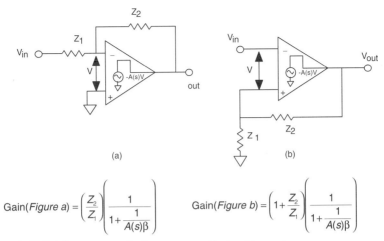

$$\text{Gain}(\textit{Figure a}) = \left(\frac{Z_2}{Z_1}\right)\left(\frac{1}{1+\frac{1}{A(s)\beta}}\right) \qquad \text{Gain}(\textit{Figure b}) = \left(1+\frac{Z_2}{Z_1}\right)\left(\frac{1}{1+\frac{1}{A(s)\beta}}\right)$$

FIGURE 6.5 Voltage feedback in op-amps with closed-loop gain.

dimensionless quantity and is expressed in decibels in source cases, but usually as a plain number (100,000 is typical). Because of its tremendous gain op-amps are not useful in the open-loop mode since a small input voltage can quickly make the op-amp to saturate, producing an output $V_{\text{out}} = V_{\text{cc}}$. The most useful configuration of course is the closed-loop configuration when a negative feedback from V_{out} is fed back to the inverting input (−) using a feedback network as shown in Figure 6.5.

The *feedback* factor is the ratio up to output signal to the signal feedback into inverting input of the amplifier and is given by

$$\beta = \frac{Z_1}{Z_1 + Z_2} \tag{6.11}$$

If we invert the feedback factor β we obtain the noise gain and it represents the voltage gain experienced by a noise voltage in series with the op-amp input terminal.

$$\text{noise gain} = \frac{1}{\beta} = 1 + \frac{Z_2}{Z_1} \tag{6.12}$$

The open-loop gain is given by

$$\text{loop gain} = A(s)\beta \tag{6.13}$$

The closed-loop gain for the inverting case is given by:

$$\text{signal gain} = -\frac{Z_2}{Z_1}\left(\frac{1}{1+A(s)\beta}\right) \tag{6.14}$$

Notice that as $A(s)$ increases (e.g., $A(s) = \infty$) the signal gain approaches $-Z_2/Z_1$. The value of $A(s)$ at a given frequency will dictate the accuracy of the op-amp of that particular frequency. As the frequency deviates from that optimum frequency (known as the corner frequency f_c), if we multiply the noise spectral density by the square root of the noise bandwidth we can then obtain the total rms noise. The most practical way to reduce the thermal noise is to minimize the bandwidth when possible.

Shot Noise: Shot noise is the noise caused by the quantized and random nature of current flow. The spectral density of the shot noise is defined by

$$|I_n|^2 = 2qI_{dc}B \text{ (Hz)} \qquad (6.15)$$

where q is the electron charge (1.6×10^{-19} C) and I_{dc} is the dc current.

Flicker Noise: The flicker noise is caused by the contamination and other defects in the silicon lattice. The combination and recombination of carriers in the emitter base of the transistor also result in the flicker noise. This process and noise are not only associated with bipolar transistor but also with CMOS processes. The current spectral density of flicker noises is given by

$$I_{nf}^2 = \frac{2qI_{dc}^m f_c B \text{ (Hz)}}{f \text{ (Hz)}} \qquad (6.16)$$

where q is the electron charge, I_{dc} is the dc current, f is the frequency of interest, and f_c is the corner frequency exponent between 1 and 2.

Popcorn Noise: Popcorn noise is also known as burst noise. It carries this name because of the noise it makes under noise amplification in an audio system. This kind of noise bursts at random amplitudes and durations. Popcorn noise is found in the low-frequency mode. This kind of noise can be described as punchthrough of emitter base junction and contamination in the emitter base region by metalization. The popcorn noise spectral density is given by

$$I_{np} = \frac{KI_c B \text{ (Hz)}}{1 + (f/f_c)^2} \qquad (6.17)$$

where K is a constant for a given device, I_c is a direct current, f_c is a corner frequency, f is the frequency of interest, and B (in hertz) is the noise bandwidth.

A sample of noise in an op-amp is shown in Figure 6.6. This model uses a noise voltage source which is in series with the noninverting input and two noise current sources between each input and ground. All these noise sources are uncorrelated.

Notice that a designer not only needs to be concerned with a specified voltage noise (given in data sheets) but also the contributing to the total noise depending on the kind of op-amp and source resistance (R_s). The total noise voltage (V_{nt}) would then be given by

$$V_{nt} = \sqrt{V_n^2 + (I_n R_s)^2} \qquad (6.18)$$

In the linearity of the op-amp, the spectral gain is shown in Figure 6.7.

Another typical and new type of op-amp is the current feedback (or transimpedance) amplifier, which is very useful at high frequencies. An equivalent

FUNDAMENTAL OP-AMP SPECIFICATIONS

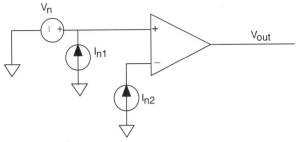

FIGURE 6.6 Noise models for a general operational amplifier.

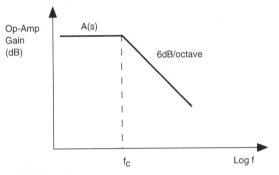

FIGURE 6.7 Spectral gain of a typical op-amp.

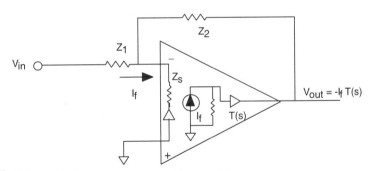

FIGURE 6.8 Inverting current feedback amplifier.

circuit is shown in Figure 6.8. It differs from the previously outlined voltage feedback in that the two inputs are addressed differently. The noninverting input (+) is a high-impedance node, as it is in a voltage feedback op-amp, but the inverting input is a low-impedance current input node. The signal from the noninverting input goes to the inverting input through a unity gain buffer. The input impedance Z_s is only a few ohms with an offset between the inverting and noninverting input; which is also an offset voltage as in a voltage feedback op-amp (hundreds of millivolts).

The following are important terms:

Inverting Case

$$\text{signal gain} = -\frac{Z_2}{Z_1}\left(\frac{1}{1+1/\text{LG}}\right)$$

Noninverting Case

$$\text{signal gain} = \left(1+\frac{Z_2}{Z_1}\right)\left(\frac{1}{1+1/\text{LG}}\right)$$

where

$$\text{LG} = \text{loop gain} = \frac{T(s)\{Z_s \parallel Z_1\}}{Z_s\{Z_s \parallel (Z_1+Z_2)\}}$$

and $T(s)$ is the open-loop transimpedance.

A comparison of the transfer functions between the voltage feedback amplifier and current feedback amplifier is shown in Figure 6.9.

Voltage Feedback

$$\frac{V_{\text{out}}}{V_{\text{in}}} = \frac{-Z_2/Z_1}{1+1/A(s)[1+Z_2/Z_1]}$$

Current Feedback

$$\frac{V_{\text{out}}}{V_{\text{in}}} = \frac{-Z_2/Z_1}{1+Z_2/T(s)[1+Z_s/Z_1+Z_s/Z_2]}$$

In voltage feedback amplifiers, the closed-loop bandwidth is inversely proportional to the noise gain, the product of the noise gain, and the closed-loop bandwidth is a constant. Therefore, the main characteristic is a constant gain bandwidth product. In current feedback amplifiers if $R_s \ll R_1$ and $R_s \ll R_2$ the preceding expression becomes:

$$\frac{V_{\text{out}}}{V_{\text{in}}} = \frac{-Z_2/Z_1}{1+Z_2/T(s)} \tag{6.19}$$

which means that the closed-loop bandwidth is very much independent of the gain Z_2/Z_1 and depends only on the feedback Z_2. It is usually appropriate for current feedback amplifiers to be optimized for maximum bandwidth with a

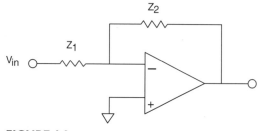

FIGURE 6.9 Comparison between voltage feedback and current feedback op-amps.

given value of Z_2. The closed-loop bandwidth of a current feedback amplifier remains constant in spite of the closed-loop gain value provided the gain is changed by varying only Z_1.

6.3. INPUT OFFSET VOLTAGE

In practical terms, to get a 0-V output a small differential voltage must be applied to the inputs. This is known as the "offset" voltage V_{os}. The V_{os} voltage can be represented as a voltage source in series with the inverting input terminal of the op-amp, as shown in Figure 6.10.

In the same manner, ideally no current should flow into the input terminals of a voltage feedback op-amp. In practice, however, there is always a bias current I_b, as shown in Figure 6.11. The value of I_b can vary from fantoamps to microamps.

Bias current can be a problem for op-amps users because this small current flow in impedances can cause voltage drops that can cause errors within the op-amp in the form of several millivolts. If the designer does not use I_b and capacitive coupling is used, the circuit may not work at all. Finally, for some op-amps (e.g., FETs) I_b varies sharply by doubling with every 10°C rise in temperature.

To reflect all the offset and bias errors to the output of the op-amp, Figure 6.12 must be used.

$$V_{out} = \pm V_{os}\left[1 + \frac{Z_f}{Z_1}\right] \pm I_{bt}Z_2\left[1 + \frac{Z_f}{Z_1}\right] \pm I_{b-}Z_f \qquad (6.20)$$

Normally the input impedance of a voltage feedback amplifier is very high, of the order of 100 MΩ. The output impedance is low and of the order of 0 to 100 Ω.

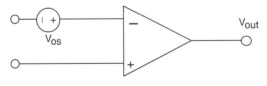

$V_{out} = (1 + Z_2/Z_1)V_{os}$

FIGURE 6.10 Offset voltage representation in an op-amp.

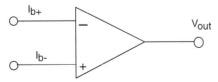

FIGURE 6.11 Input bias current representation.

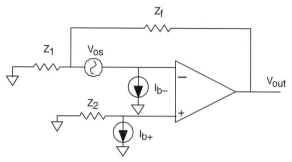

FIGURE 6.12 Integrated bias and offset voltage representations.

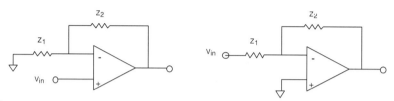

FIGURE 6.13 Noise gain in voltage feedback amplifiers.

6.4. THE NOISE GAIN OF OP-AMPS

The noise gain of a voltage feedback amplifier is dependent on the feedback configuration, as shown in Figure 6.13.

$$\text{noise gain} = 1 + \frac{Z_2}{Z_1} \qquad \text{noise gain} = 1 + \frac{Z_2}{Z_1}$$

$$\text{signal gain} = 1 + \frac{Z_2}{Z_1} \qquad \text{signal gain} = -\frac{Z_2}{Z_1}$$

The expression for a closed-loop gain G is dependent on the noise gain G_n and the open-loop gain of the amplifier A and is given by:

$$G = \frac{V_{\text{out}}}{V_{\text{in}}} = \frac{AG_n}{G_n tA} \qquad (6.21)$$

6.5. SLEW RATE AND POWER BANDWIDTH OF OP-AMPS

The slew rate of an amplifier (see Figure 6.14) is defined as the maximum rate of change of voltage at the output, and it can be expressed as volts per microseconds.

$$V_{\text{out}} \cong -\frac{Z_2}{Z_1} \frac{V_0}{2} \sin 2\pi ft \qquad (6.22)$$

where V_0 is the peak-to-peak input voltage. The slew maximum slew rate is given by

$$\text{slew rate} = \left.\frac{dv}{dt}\right|_{\text{max}} = -\frac{Z_2 V_0}{2Z_1}(2\pi f) = \frac{-Z_2 V_0 \pi f}{Z_1} \qquad (6.23)$$

GAIN–BANDWIDTH PRODUCTS

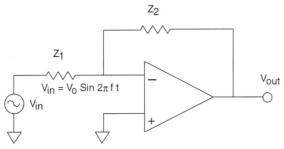

FIGURE 6.14 Diagram for slew rate representation in op-amps.

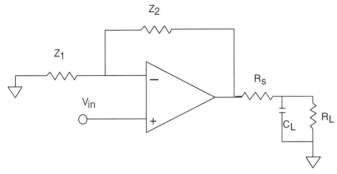

FIGURE 6.15 Proper driving of capacitive loads.

The full-power bandwidth (FPBW) of an op-amp is the maximum frequency at which slew limiting does not occur at maximum output on

$$f = \text{FPBW} = \frac{\text{slew rate}}{(-Z_2 V_0 \pi)/Z_1} \quad (6.24)$$

The slew rate can also depend on the power supply voltage and the load the amplifier is driving. When the load is capacitive, its output voltage can be slowed down, which can cause instability in the op-amp due to its negative feedback. When an op-amp drives a capacitive load a series resistor (R_s) must be used outside the negative feedback loop as shown in Figure 6.15. This method, however, reduces the total bandwidth of the op-amp.

6.6. GAIN–BANDWIDTH PRODUCTS

The open-loop gain (in decibels) of a voltage feedback amplifier with only one pole is shown in Figure 6.16.

The closed-loop bandwidth (f_{CLBW}) is the frequency at which the noise gain intersects the open-loop gain. The gain–bandwidth product (GBW) is given by

$$\text{GBW} + (\text{noise gain})(f_{\text{CLBW}}) \quad (6.25)$$

$$\text{GBW} = \left(1 + \frac{Z_2}{Z_1}\right) f_{\text{CLBW}} \quad (6.26)$$

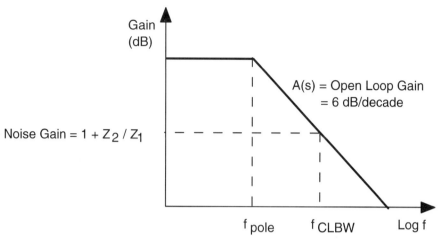

FIGURE 6.16 Open-loop gain of an op-amp.

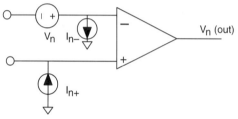

FIGURE 6.17 Representation of intrinsic input noises.

6.7. INTERNAL NOISE IN OP-AMPS

Noise can enter an amplifier in the form of input voltage noise or input current noise, as shown in Figure 6.17. The input noise V_n is bandwidth dependent and measured in $nV/\sqrt{Hz}$, which is its spectral density. In voltage feedback op-amps, the noise current (I_{n-}, I_{n+}) in the inverting and noninverting inputs is uncorrelated and about equal in magnitude. In BJT and FET input stages, the noise current is the shot noise of the bias current and can be obtained from the bias current.

Noise current becomes important when the source impedance is significant such that the induced noise voltage is greater than the thermal noise or voltage noise. The choice of a low-noise op-amp depends on the source impedance of the signal and when the source impedance is high, carried noise usually dominates. Therefore, we should carefully consider the kind of amplifier to use dependent on the impedance circuitry used with the amplifier, as shown in Figure 6.18.

The noise figure of an operational amplifier in a given circuit is the amount by which the noise of the circuit exceeds that of the same circuit when using a noise-free amplifier. At low frequencies, the noise spectral density goes up at 3 dB/octave, as shown in Figure 6.19. The frequency at which it starts to rise is known as the $1/f$ corner frequency. In the $1/f$ region, the rms noise in the

INTERNAL NOISE IN OP-AMPS

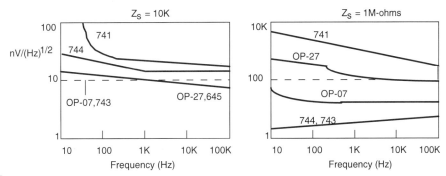

FIGURE 6.18 The effect of source impedance variations vs frequency for different amplifiers.

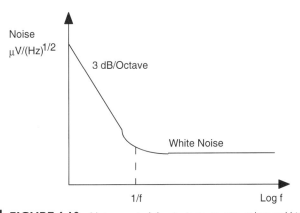

FIGURE 6.19 Noise spectral density in an op-amp at low and high frequencies.

bandwidth $\Delta f = f_1 - f_2$ is given by

$$V_{\rm rms} = k\sqrt{\ln\left(\frac{f_2}{f_1}\right)} \qquad (6.27)$$

where k is the noise spectral density at 1 Hz.

To analyze the noise performance of an op-amp, the noise contribution from each part of the circuit must be assessed, as can be observed in Figure 6.20, where we have added the capacitors C_1 and C_2 to a voltage feedback op-amp. On the inverting input of the op-amp C_1 is the sum of the op-amp internal capacitance and any other external feedback capacitance, and C_2 is a feedback capacitor for stabilizing the circuit. The noise voltages V_{zs}, V_{z1}, and V_{z2} are thermal Johnson voltages of the types $\sqrt{4KTZ_s}$, $\sqrt{4KTZ_1}$, and $\sqrt{4KTZ_2}$, respectively. The terms I_{n+} and I_{n-} are the intrinsic noise currents of the amplifier. Finally, V_n is the intrinsic noise voltage of the amplifier.

The calculation of the total output rms noise is made by multiplying each of the noise voltages in Figure 6.20 by the gain and integrating over the frequency range of interest, as shown in Table 6.1.

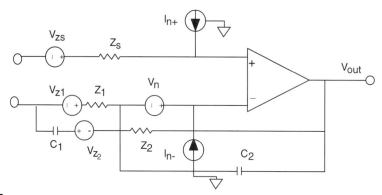

FIGURE 6.20 Noise contributions from passive and active circuit elements in an op-amp.

TABLE 6.1 Noise sources and calculation of total output noise

Noise source expressed as a voltage	Multiplication by this factor reflected to the output	Integration bandwidth
$\sqrt{4KTR_s}$	Noise gain as a function of frequency	Closed-loop bandwidth
$(I_{n+})Z_s$	Noise gain as a function of frequency	Closed-loop bandwidth
V_n	Noise gain as a function of frequency	Closed-loop bandwidth
$\sqrt{4KTR_s}$	$-Z_2/Z_1$ (signal gain)	$\frac{1}{2}\pi Z_2 C_2$ (signal bandwidth)
$\sqrt{4KTR_s}$	$-Z_2/Z_1$ (signal gain)	$\frac{1}{2}\pi Z_2 C_2$ (signal bandwidth)
$(I_{n-})Z_2$	$-Z_2/Z_1$ (signal gain)	$\frac{1}{2}\pi Z_2 C_2$ (signal bandwidth)

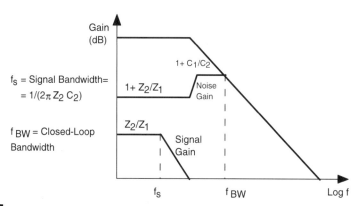

FIGURE 6.21 Noise gain in inverting feedback amplifier.

The sum of all the root sum square contributions will contribute to the total output noise. The noise gain is shown in Figure 6.21. From Figure 6.21 we can observe that the output noise resulting from the input noise voltage is determined mainly by the high-frequency portion where the noise gain is given by $1 + C_1/C_2$.

6.8. NOISE ISSUES IN HIGH-SPEED ADC APPLICATIONS

In high-speed and wide-bandwidth analog-to-digital converters, when a wide range of analog input frequencies is used, a wide range of outputs is usually produced. This is the result of the front-end and wide-bandwidth and other noise concerns. The correct output shows most of the time but adjacent outputs also appear, though with reduced probability. This scenario is as shown in Figure 6.22.

In the process of driving analog-to-digital converters with wide-bandwidth op-amps, the output noise of the amplifier driver will contribute to the overall analog-to-digital converter noise floor. Therefore, the analog-to-digital converter noise should always be compared with the front-end wideband amplifiers. The noise model for the op-amp is shown in Figure 6.23.

$$V_{out} = \sqrt{BW} \left[I_n^2 Z_2^2 + I_{n+}^2 Z_3^2 \left[1 + \frac{Z_2}{Z_1} \right]^2 + V_n^2 \left[1 + \frac{Z_2}{Z_1} \right] \right.$$

$$\left. + 4KTZ_2 + 4KTZ_1 \left[\frac{Z_2}{Z_1} \right]^2 + 4KTZ_3 \left[1 + \frac{Z_2}{Z_1} \right]^2 \right]^{1/2} \quad (6.28)$$

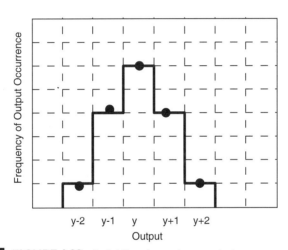

FIGURE 6.22 Probability of several outputs in the presence of noise.

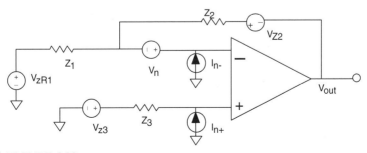

FIGURE 6.23 Noise model of an op-amp.

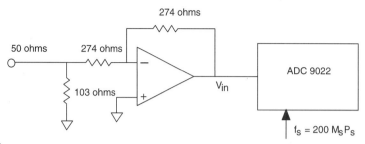

FIGURE 6.24 Example of noise calculation using a wideband input amplifier.

where $BW = 1.57 f_{BW}$. The 1.57 is required to convert the single-pole ADC input into an equivalent noise bandwidth, and f_{BW} is the closed-loop bandwidth.

The Johnson noise contribution of the resistor in Eq. (6.28) (e.g., $4KTZ_2$) can often be neglected if the source resistance is less than 1 KΩ, and this is often true for high-speed systems. In voltage feedback op-amps, the input current noise (I_{n+}, I_{n-}) can usually be neglected. In current feedback op-amps, the inverting input current noise (I_{n-}) usually dominates; at higher gains, however, the voltage noise V_n becomes significant. An example of noise calculations using a wideband input amplifier into an ADC is shown in Figure 6.24.

Specification Data

> AD9632 op-amp input noise specs: 4.3 nV/$\sqrt{Hz}$
> Closed-loop bandwidth BW = 250 MHz
> AD9022 effective input noise specs: 285 μV rms
> Input bandwidth 110 MHz

Solution

> AD9632 noise output spectral density = $2 \times 4.3 \, nV/\sqrt{Hz} = 8.6 \, nV/\sqrt{Hz}$
> $V_{in} = 8.6 \, nV/\sqrt{Hz} \sqrt{(110 \times 10^6 \, Hz)(1.57 \, Hz)} = 113 \, \mu V \, rms$

Notice that 113 μV rms is less than 285 μV rms. Therefore, we should not be concerned with the wideband op-amp noise affecting the analog-to-digital converter output data. The bandwidth for integration should be the lower of either the ADC bandwidth or the op-amp bandwidth.

In most high-speed analog-to-digital converter system applications, a passive antialiasing filter (low pass for baseband sampling or bandpass for harmonic sampling) is used between the wideband op-amp and the analog-to-digital converter, as shown in Figure 6.25. This approach will reduce even more the effects of the driver wideband amplifier noise. The op-amp noise rarely interferes with the performance of high-speed digital-to-analog converter systems. The major contributors to noise in such designs are (1) power grounding and layout, (2) poor decoupling methods, (3) noisy clock, and (4) external switching power supply.

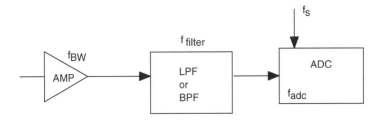

$f_{filter} < f_s/2 \ll f_{adc} < f_{BW}$

FIGURE 6.25 Use of low-pass filter to reduce wideband amplifier noise.

6.9. PROPER POWER SUPPLY DECOUPLING

Good power supply decoupling methodologies must be used on each and every circuit board. The need for power supply bypass results from the parasitic impedances in supply lines, which degrade the noise and stability performance. The supply line impedances, mostly of inductive nature, which supply the amplifier current and other circuits that feed from the same power lines, provide the power supply noise current and voltages. The supply current flows through these inductive impedances, producing voltage drops at the amplifier supply connections and interpreted as noise voltages. This voltage noise causes the amplifier to activate its power supply rejection ratio, reproducing a portion of the noise voltage at the op-amp inputs. The coupled noise combines with the internal noise of the amplifier and then is amplified by the noise gain of the amplification circuit. The portion of the input noise from the amplifier itself constitutes a parasitic feedback signal that could eventually produce oscillations.

In an effort to minimize inductance, wide bus traces could be used. Furthermore, minimizing bus lengths could also help. However, the most significant contributor is the use of capacitive bypass of the power supply lines. This approach, if properly performed, will frequently ensure stability by minimizing interference—the interconnection length and the associated inductance between the amplifiers and the capacitors (surface-mounted capacitors work best). Sizing the bypass capacitors is also a consideration since too large a capacitor can cause internal parasitic impedances to be added to the capacitor lead impedances.

In Figure 6.26 we show how the power supply line couples noise into the operational amplifier. In the figure, the voltages V_{s+} and V_{s-} provide the bias voltages for the positive and negative power supplies in the operational amplifier. Under ideal conditions, V_{s+} and V_{s-} correspond to the supply voltages' positive and negative levels. However, the supply line inductances L_s react with the supply current I_s, producing voltage potential drops along the power supply lines. The supply current I_s is the sum of the supply current drawn by the op-amp of Figure 6.26 and other circuitry powered from the same power supply lines. The L_s inductance accounts for all inherent parasitic inductances that may exist, including trace layout (15 nH/in.). The more complex the PCB and the power supply wiring, the more inductance we will have, and, in a complex

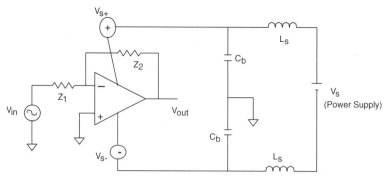

C_b = Bypass Capacitance (1.0 through 10 μF)
L_s − Series Inductance (100 nH through 1000 nH)

FIGURE 6.26 Coupled noise into op-amp from power supply rails.

matter L_s can reach several hundreds of nanohenries. In the figure, we see the addition of bypass capacitors C_b.

The bypass capacitor shunts the line impedances to reduce the supply line voltage drops produced by I_s. Therefore, the bypass capacitors C_b attenuate the supply coupling effects. From an electronics point of view, C_b also serves the main and immediate source of high-frequency current needs. Otherwise, such current demands from the op-amp would require serious time delays in their travel from the power supply to the op-amp. The delay would cause phase shifts in the amplifier response, which usually increases with frequency. The usefulness of the bypass capacitor is that it supplies much of the high-frequency current demand, eliminating the time delay and the corresponding phase shift.

Though bypass capacitors are of great help, they do not completely eliminate the power supply coupling problems. A voltage difference still develops with the trace-line impedance Z_0 and the supply current I_s, reducing the voltage magnitude at both V_{s+} and V_{s-} by the amount of $I_s Z_0$.

$$V_{s+} = V_s - I_s Z_0 \tag{6.29}$$

$$V_{s-} = V_s + I_s Z_0 \tag{6.30}$$

Therefore the total supply voltage delivered to the op-amp is given by $V_+ - V_- = V_s - 2I_s Z_0$, which means the supply voltage decreases by $2I_s Z_0$. The decrease causes an activation of the power supply rejection ratio (PSRR) of the op-amp, producing an amplifier error at its input of magnitude given by $V_e = 2I_s Z_0 / \text{PSRR}$. The op-amp amplifies this error signal with the noise gain $A_n(s)$ that amplifies the op-amp's input noise voltage. The output noise voltage is given by

$$V_{\text{out}}(\text{noise}) = \frac{2A_n(s) I_s Z_0}{\text{PSRR}} \tag{6.31}$$

where

$$A_n(s) = \frac{A(s)}{1 + A\beta} \tag{6.32}$$

where $A(s)$ is the open-loop gain, and β is the circuit's feedback factor. For the configuration in Figure 6.26,

$$A_n \cong \frac{1 + Z_2/Z_1}{1 + f/f_c} \qquad (6.33)$$

where f_c is the unity gain crossover frequency of the op-amp.

The supply noise $V_{out}(noise)$ in Eq. (6.31) is frequency dependent such that $V_{out}(noise)$ increases as frequency increases. The supply bypass capacitor reduces the $V_{out}(noise)$ response from a double zero to the flat response, which is usually expected for op-amp noise. For the unbypassed case, a diminished PSRR and an increasing line impedance introduce zeros in the $V_{out}(noise)$ response. When bypassed a diminished PSRR and a diminished impedance produce a cancelling effect and a flat frequency response.

Another important point concerning bypass capacitor usage is that supply-line coupling produces parasitic feedback. In addition to noise reduction, the power supply bypass capacitors must try to preserve frequency stability (i.e., preventing oscillations). The bypass capacitor selection should also focus on improvements to prevent oscillation. Stability requires the use of bypass capacitance to control the parasitic feedback loop established by the voltage supply lines impedances and the amplifier's PSRR coupling. In Figure 6.27, a $V_{out}(noise)$ voltage supplies a load current (I_L) to load Z_L. The amplifier draws this current from V_{s+} and through Z_0 impedance of its supply line. The resulting line voltage drop produces a component of the error voltage V_e, $-I_L Z_0/PSRR$, through the amplifier's finite PSRR. This circuit will amplify this component by the circuit noise gain $A_n(s)$ producing a $V_{out}(noise)$ output response. This response is reflected back to the amplifier inputs through the amplifier's open-loop gain $A(s)$, creating the $V_{out}/A(s)$ component of the V_e shown.

Therefore, the power supply coupling produces an input signal that in turn would yield an output signal. This output will then produce an input signal. This scenario describes the full circle of the feedback loop, which is capable of

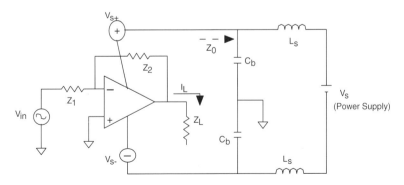

C_b = Bypass Capacitance (1.0 through 10 μF)
L_s = Series Inductance (100 nH through 1000 nH)

FIGURE 6.27 Illustration of output noise generated by the wideband op-amp.

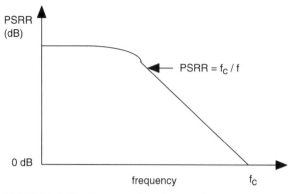

FIGURE 6.28 Illustration for obtaining the unity gain crossover frequency to avoid op-amp oscillations.

sustaining an oscillation. It can be shown that keeping Z_0 low such that

$$Z_0 < \frac{Z_L \text{PSRR}}{A_n(s)} \quad (6.34)$$

where

$$A_n(s) = (1 + Z_2/Z_1)/(1 + f/f_c)$$

avoids oscillation, where f_c is the unity gain crossover frequency of the op-amp and is shown in Figure 6.28.

6.10. BYPASS CAPACITORS AND RESONANCES

Adding a power supply bypass capacitor produces two LC resonances. The line inductance of the power supply and the basic bypass capacitor itself produce the first resonance. The bypass capacitor itself produces the second resonance. The first resonance from C_b and L_s forms an LC network with an impedance of

$$Z_0 = \frac{sL_s}{1 + s^2 L_s C_b}$$

At lower frequencies $Z_0 \cong sL_s$ and at high frequencies $Z_0 \cong 1/sC_b$.

At an intermediate frequency, this impedance displays a resonance maximum given by

$$f_R = \frac{1}{2\pi\sqrt{L_s C_b}} \quad (6.35)$$

At such a resonance, we could imagine the impedance reaching infinity. However, this actually does not happen because of the power supply line's parasitic resistance R_s. This line resistance dissipates the resonant energy of the LC circuit causing a decrease of what could have been a large impedance rise. A typical PCB trace can dissipate 12 mW/in. of the power supply energy. The resonance frequency value should be such that it is less than f_c and at a location where

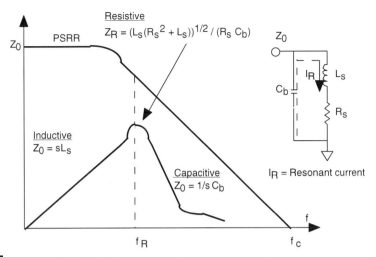

FIGURE 6.29 The first resonance frequency f_R formed by the power supply line inductance and bypass capacitance.

PSRR is higher in order for the amplifier to be able to attenuate the coupling effects. This can be observed in Figure 6.29. In the figure, we can observe that the power supply line impedance varies from inductive to resistive to capacitive as the frequency goes up. At low frequencies L_s predominates in Z_0, producing an upward slope in the impedance curve. At high frequencies, the bypass capacitor C_b takes over producing downward slope in the impedance curve. Between these two frequency ranges, the induction and capacitive slopes intersect at a resonance frequency f_R. At f_R, the phase difference between the upward and downward slope curves transforms the equal magnitude in both curves into resonance. The resonant current I_R results from the oscillation caused by the energy transfer from the inductor L_s to the capacitor C_b and vice versa. The analysis of the LRC circuit in Figure 6.29 shows that

$$Z_{0R} = \frac{\sqrt{L_s(R_s^2 C_b + L_s)}}{C_b R_s} \qquad (6.36)$$

By making C_b large, we can decrease the value of the Z_{0R} impedance to a value given by

$$Z_{0R} = \sqrt{\frac{L_s}{C_b}} \qquad (6.37)$$

It was previously stated that the value of the resonance frequency should be such that

$$f_R = \frac{1}{2\pi\sqrt{L_s C_b}} \ll f_c \qquad (6.38)$$

which implies that

$$C_b \gg \frac{1}{4L_s(\pi f_c)^2} \qquad (6.39)$$

It has been shown that a general design equation for C_b is given by the expression

$$C_b = \frac{50}{\pi f_c} \qquad (6.40)$$

It was previously stated that the bypass capacitor itself introduces a second resonance. The inherent parasitic inductance and resistance of capacitors can also disturb the bypass capabilities. The inductance will introduce a new resonance, and the resistance will limit the line impedance reduction. At this new resonance frequency, the bypass impedance would drop to zero, except that, as before, we also have a parasitic resistance that prevents this from happening. Above the resonant frequency, the capacitor's parasitic inductance overrides the capacitance. Large capacitors tend to introduce a new resonant condition that comprises the bypass effectiveness at frequencies typically within the amplifier's response range. This new resonance results from the parasitic inductance L_b. All capacitors pass this internal inductance. It depends on the capacitors' internal conductive paths and leads. Reducing the total connecting length can diminish the parasitic inductance. This can be accomplished by minimizing the capacitor lead length, circuit board traces, and internal path components.

A detailed examination of the bypass capacitor's actual impedance is shown in Figure 6.30. The capacitor parasitic inductance appears in series with the intended capacitance along with a parasitic resistance. The inductance L_b is self-resonance with C_b. The parasitic resistance R_b sets the capacitor impedance at resonance. This resistance comes from the same connecting path that produces the capacitor's inductance. The parasitic resistance R_b causes a voltage drop that limits the impedance decline caused by the $C_b - L_b$ resonance. This parasitic impedance detunes the $C_b - L_b$ resonance, decreasing what could have been a large phase transition in the power supply line impedance. The transition presents a broad range of phase conditions that could degrade stability.

The resistance benefits the performance of the bypass capacitor as long as it delivers the resonance. The resonance frequency at which the L_b and C_b impedances become equal is given by

$$f_{R_b} = \frac{1}{2\pi \sqrt{L_b C_{bc}}} \qquad (6.41)$$

The minimum resistance required for detuning L_b and C_b is given by

$$R_b = \sqrt{\frac{L_b}{C_{bc}}} \qquad (6.42)$$

The design limit for R_b is given by

$$\sqrt{\frac{L_b}{C_{bc}}} \leq R_b \leq 1 \qquad (6.43)$$

We can now summarize the results of Figures 6.29 and 6.30 and develop a single plot of a single bypass capacitor behavior, which is shown in Figure 6.31.

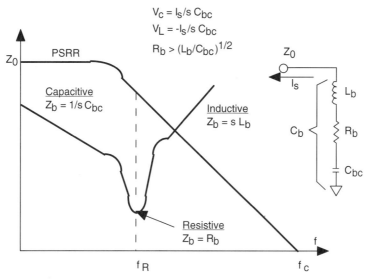

FIGURE 6.30 Analysis of bypass capacitor second resonance.

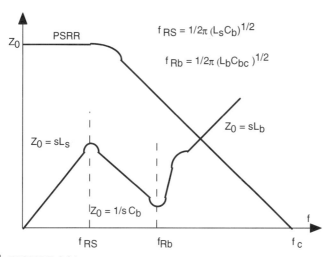

FIGURE 6.31 Composite resonance study of a single bypass capacitor behavior.

In this case, a single capacitor can bypass the parasitic inductance L_s of the power supply. For this to be effective, the capacitor must be placed very close to the operational amplifier supply lines in order to bypass the power supply inductance. The supply inductance dominates the line impedance Z_0 at lower frequency, which increases proportionally with the line inductance ($Z_0 \cong sL_s$). At somewhat higher frequencies the bypass capacitance reverses the slope of this response temporarily. First, the impedance response goes from $Z_0 = sL_s$ to $Z_0 = 1/sC_b$, with capacitive shunting diminishing the effect of the L_s inductance. As the frequency increases even further Z_0 starts rising again as the capacitor's own parasitic inductance overrides the capacitive shunting and the impedance becomes $Z_0 = sL_b$, where L_b is the bypass capacitor's parasitic inductance. The resonances f_{R_s} and f_{R_b} indicate the transition points separating the three Z_0 regions. It is important to carefully select the kind of bypass capacitor to be used. The bypass capacitor must bypass the Z_0 impedance over the entire amplifier response range. If we pick a capacitor that is quite large with the purpose of reducing the net line impedance, f_{R_s} will move to the left in the lower frequency range but the capacitance's own parasitic inductance L_b will increase causing the f_{R_b} frequency to be moved further to the right. A compromise can be reached by letting

$$C_b = \frac{50}{\pi f_c} \qquad (6.44)$$

6.11. USAGE OF TWO OR MORE BYPASS CAPACITORS

More op-amps these days use a wider frequency range, and greater amplifier bandwidth covers more of the high frequency. Due to the higher frequency requirements of these op-amps, a second bypass capacitor may be needed to counter the inductance of the primary bypass capacitor. When we add a smaller capacitor in parallel with the first bypass capacitor the inductance limit of the first capacitor is bypassed. However, the second capacitor also has an inductance of its own producing another bypass scenario at a higher frequency. Furthermore, the inductance of the first capacitor provides a resonance when combined with the second capacitor.

Adding a second bypass capacitor in parallel with the first capacitor provides a low-bypass impedance for the full response range of the high-frequency amplifier. The first capacitor $C_{b1} \gg C_{b2}$ (second capacitor). The lower capacitance of C_{b2} and its lower parasitic inductance produce a higher resonance frequency. Therefore, when we add a second bypass capacitor, we restore the declining frequency of the bypass impedance but there are also some minor complications with the introduction of two additional resonances; one from the self-resonance of the second capacitor and the other one from the interaction between the second with the inductance of the first capacitor.

In Figure 6.32 we see an illustration of the new resonance with a circuit model and the corresponding impedance responses. There are basically two corners in Figure 6.32 representing the capacitor impedances Z_{b1} and Z_{b2}. At lower frequencies, a declining Z_{b1} (i.e., from C_{b1}) provides the lower impedance bypass shunt. At a higher frequency Z_{b1} starts resonating and begins to rise at $f_{R_{b1}}$. As the frequency increases even further, Z_{b2} (i.e., from C_{b2}) bypasses the rise and restores a declined bypass impedance. The self-resonance of C_{b2} at $f_{R_{b2}}$ produces a rise but at a lower impedance than provided by Z_{Cb1}. In Figure 6.32 the C_{b1}/C_{b2} parallel setup peaks at f_{ib}, which is the intercept of the rising Z_{b1} curve and the falling Z_{b2}. For higher-frequency amplifiers this peak should fall within the amplifier's response range. At the f_{ib} intercept point the two Z_{b1}, Z_{b2} curves occupy the same value ($Z_{b1} = Z_{b2}$). At this point also $Z_{b1} = 2\pi f_{ib} L_{bi}$ and the capacitance impedances $Z_{b2} = 1/2\pi f_{ib} L_{b1}$ and the capacitance impedance $Z_{b2} = 1/2\pi f_{ib} C_{b2}$ are equal, which means that equating

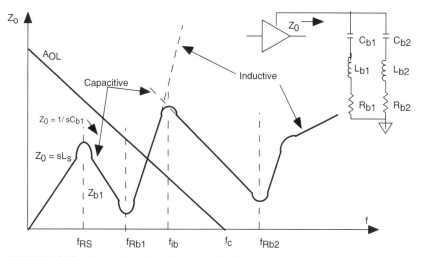
FIGURE 6.32 New and combined resonance for bypass capacitor in an op-amp.

these two terms we have

$$f_{ib} = \frac{1}{2\pi\sqrt{L_{b1}C_{b2}}} \quad (6.45)$$

Furthermore at f_{ib} the Z_{b2} impedance continues its capacitance rolloff, as shown by $Z_{b2} = 1/2\pi f_{ib}C_{b2}$. This will result in the following design equation

$$|C_{b2}| = |L_{b1}| \quad (6.46)$$

Therefore, making the magnitude of the C_{b2} capacitance equal to that of C_{b1} parasitic inductance. This design equation requires the measurement of L_{b1}, which is not a hard task to do with the accuracy of today's impedance analyzers, which can measure the frequency response of a given capacitor. The rising part of this impedance curve will define the actual inductance by using the equation $L_{b1} = Z_{\text{cap}}/2\pi f$.

The dual-bypass configuration can produce a critical resonance that degrades stability at certain frequencies. This can occur at frequencies that could be located either above or below the amplifier's crossover frequency f_c. The C_{b2} resonance in conjunction with L_{b1} can raise the net line impedance well above this level, producing oscillations. These resonances can provoke oscillation at frequencies above f_c, which can diminish the parasitic feedback loop, but the resonance impedance rise can counteract this limit. Resistive detuning of the bypass impedance can detune this resonance also. Adding a small resistance series with C_{b1} detunes this resonance to ensure stability, as shown in Figure 6.33. In the figure, the Z_0 curve now makes a slow, rather than resonant, transition between Z_{b1} and Z_{b2} at the f_{ib} intercept. The addition of a resistance R_s actually detunes these two resonances. The first resonance to be detuned is the self-resonance of C_{b1} and then the resonance from C_{b1} and C_{b2} combined. The addition of R_s removes the resonance impedance drop and raises the impedance level to that of $R_s + R_b$. This raises the bypass impedance in the region that previously was that of f_{ib}. The Z_0 curve makes a smooth, rather

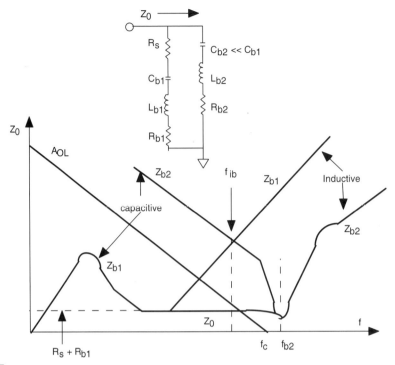

FIGURE 6.33 The effect of adding a small resistance for detuning main resonance in op-amp bypass capacitance usage.

than resonant transition to this new limit level. The reduced Z_0 response slope provides a greatly reduced phase transition at the frequency of the previous f_{b1} resonance.

This scenario reduces the potential phase combinations with amplifier gain in PSRR that could degrade stability. The value of R_s has been defined as

$$R_s = 1 - R_{b1} \tag{6.47}$$

where R_{b1} is really a very small resistor and we may choose a different capacitor for C_{b1}, which may contain this parasitic resistance (i.e., $R_s + R_{b1}$).

6.12. DESIGNING POWER BUS RAILS IN POWER-GROUND PLANES FOR NOISE CONTROL

Providing power for microprocessor-based systems is becoming an increasingly difficult job for advanced digital designs. The reason is that power supply rails are dropping in voltage over the years; from 5 to 3.3 V and to 1.0 V in the future. From the IC process, lithography demands lower, better regulated power supply rails and higher clock speeds. This creates noise and dynamic loads' ever-increasing demand for lower power contribute to power design problems.

Power supply rails for microprocessors and other high-speed ICs have dropped from 5 V to about 3.3 V in order to handle 0.5-μm lithography and reduce total power consumption. The goal in the near future is to drop to 1 V for 0.35-μm lithography, which would reduce the power consumption even further. Since 1985, clock rates have risen about 25% per year. If this trend continues by the year 2002 the clock speed could easily reach 400–1200 MHz. Clocks that are just 50 MHz will convert a supply rail into a transmission line by raising the supply's source impedance and radiating RF noise. It is not enough that IC's basic clock rate switching noise rides on the supply rail, but power management schemes produce near maximum di/dt load current transients in just one clock cycle.

When ICs and microprocessors come out of sleep mode, load transients are easily developed even for low current such as 3–4.0 mA in a typical microprocessor. Such a processor can see 300-A/μs transients (3-A rise in a 10-ns clock cycle). A typical PCB trace can look like a 20 nH/in. in inductance as shown in Figure 6.34. Because the voltage drop across the transmission line equals the inductance multiplied by the rate of change of current ($V = L di/dt$), the results are given by

$$V = 20.0 \times 10^{-9} \times 300\,\text{A}/1.0 \times 10^{-6} = 6\,\text{V/in.}$$

which means that if the processor is located about 1 in. from a typical 3.3-V power source, the voltage at its pins will drop to zero when called out of the sleep mode.

Digital switching noise on the supply rail will see skin effects that will produce voltage drops even when the dc resistance is quite low. Furthermore, the RF noise riding on the 3-V rail can couple back through the dc–dc converter to the power source.

The task facing the power and ground designers of printed circuit boards is to provide the shortest possible path for the return of common mode noise currents. Though noise currents always follow the path of least impedance, in the frequency range of 10 kHz to 50 MHz it is often difficult to predict which path the noise current will follow. As a general rule, however, reducing the inductance and increasing the capacitance of the loop through which the current will follow reduce the overall impedance. The inductance of a wire

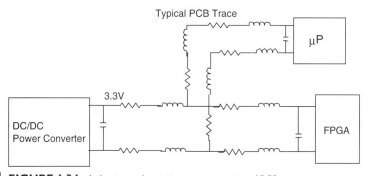

FIGURE 6.34 Inductive and resistive representation of PCB power traces.

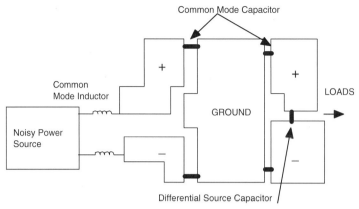

FIGURE 6.35 Capacitive and inductive placement for routing away conducted noise in PCB.

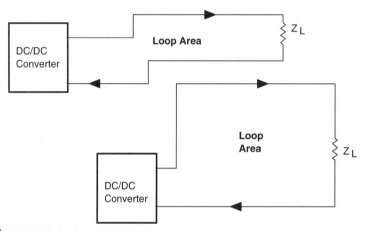

FIGURE 6.36 Reducing loop area in PCB for diminishing interference.

loop is proportional to the area of the loop. The smaller the loop area the smaller the inductance and the greater is the capacitance and the lower the overall impedance. Capacitors in the range of 0.01 to 0.1 µF can be placed to route conducted noise away from sensitive ICs and back toward a nonnoise return path, as shown in Figure 6.35. If more severe noise limits are imposed, a common mode inductor can be inserted into the noisiest circuit, as shown in Figure 6.35. This inductor is typically about less than 1 mH but would still provide enough common mode and differential mode filtering to reduce the conducted noise significantly.

Because the impedance of a wire loop is proportional to its area, reducing the area that is enclosed by the loop reduces the impedance. The area between traces can be minimized by placing the traces closest to each other in adjacent layers of the printed circuit board. A power bus loop area can also be reduced by decreasing the distance between the power source (e.g., a dc–dc converter) and the load. Though this is often impractical the loop area can also be reduced significantly by using a good design as shown in Figure 6.36, where the loop

THE EFFECT OF TRACE RESISTANCE

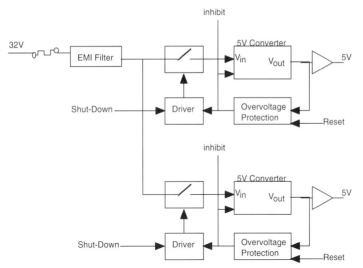

FIGURE 6.37 Block diagram of an overvoltage protection circuit for the dc–dc converter.

area has been reduced significantly and therefore reducing significantly the RF impedance. It is also less efficient as an antenna because its smaller loop area generates a smaller magnetic field.

Likewise minimizing the loop area decreases significantly the noise pickup. In power converters, not only the power leads but the two sense leads, traces must be put as close to each other as possible on opposite sides of the PCB layer. Any inductance in the loops of Figure 6.36 provides a delay in the converter control loop response and can reduce transient response to load steps, creating an unstable converter.

Not only loop reduction must be pursued in power signal lines but likewise signal lines that can control the turning in and turning off of a dc–dc converter must also be protected from coupled noise, which could adversely affect the on-off line, inhibit line, and turn the converter on and off in an inadvertent manner. In Figure 6.37 a dc–dc converter is being protected from overvoltage by the use of an overvoltage protection (OVP) circuit as shown in the block diagram of Figure 6.36. The overvoltage protection circuit consists partially of a series of flip-flops and NAND gates which generate the inhibit signal once an overvoltage is detected. Unfortunately these lines are very susceptible to EMI noise and minimizing the loop area in the design to reduce the possibility of coupled electromagnetic interference noise is highly imperative.

6.13. THE EFFECT OF TRACE RESISTANCE

Clear power at the output is dependent on the power supply capacity and its tolerance but the loss resistance between the supply and the printed circuit board output is also an important factor, as shown in Figure 6.38. The system power supply must have enough drive for any other peripherals plus additional current for the PCB card. The output leads of the converter should have a

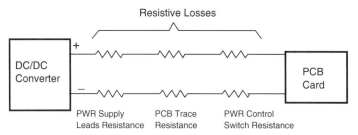

FIGURE 6.38 Trace resistance in PCB traces.

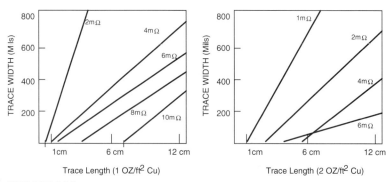

FIGURE 6.39 Trace width and length for a given thickness of copper in PCB trace manufacture.

very low resistance. Trace resistance can also be minimized by careful design. The current-handling capability of a trace is proportional to its cross-sectional area. An insufficiently large trace can act as a fuse with a high current flow, so trace cross-sectional area should be maximized. Therefore, both minimum resistance and maximum current-handling capability can be achieved with heavy traces. Figure 6.39 shows trace widths as trace length for 1- and 2-oz/ft² copper thickness.

The following three design equations provide a methodology for calculating the resistance and current density of a PCB trace as a function of heat dissipation.

$$\rho_s(T) = \frac{\rho[1 + \alpha(T_A + T_{\text{rise}} - 20)]}{h} \quad (6.48)$$

where

$\rho_s(T)$ = sheet resistance at elevated temperature (Ω/square)
$\rho = 0.0172$ = copper resistivity at 20°C ($\Omega \times \mu m$)
$\alpha = 0.00393$ = temperature coefficient of ρ (°C)
T_A = ambient temperature (°C)
T_{rise} = allowed temperature rise (°C)
h = copper trace height (μm)

$$W = \frac{100\, I_{\max}}{\sqrt{[(T_{\text{rise}}/\theta_{\text{SA}})/\rho_s(T)]}} \quad (6.49)$$

where

W = minimum copper trace width (mils)
I_{max} = maximum allowed current for a given T_{rise} (A)
T_{rise} = allowed temperature rise (°C)
θ_{SA} = trace thermal resistance (°C × in.²/Ω)
$\rho_s(T)$ = sheet resistance at elevated temperature (Ω/square)

$$I = \frac{WR}{\rho_s(T)} \qquad (6.50)$$

where

I = maximum trace length (mils)
W = trace width (mils)
R = maximum allowed resistance (Ω)
$\rho_s(T)$ = sheet resistance at elevated temperature (Ω/square)

These equations and graphs enable us to calculate the maximum trace width and the maximum trace length for a given allowed temperature rise for the trace. Using the PCB weight in Table 6.2 we can get the copper trace height. From this information and the allowed temperature rise we can first calculate the sheet resistance $\rho_s(T)$. The minimum trace width can be obtained from Eq. (6.50) if the maximum current the trace can sustain is given. Finally, based on the maximum allowed resistance, the maximum trace length is calculated using Eq. (6.51).

Table 6.2 provides general guidance based on a logic family for situations in which the loaded transmission line delay is unknown or difficult to obtain.

Depending on the logic family the characteristic source impedance can vary. For ECL, source and load impedances are about 50 Ω. For TTL, the source impedance ranges from 70 to 100 Ω. Implementation of transmission lines should try to match the logic family impedance. There are four implementations of traces: (1) microstrip, (2) embedded microstrip, (3) centered stripline, and (4) dual off-center stripline. The microstrip is the simplest configuration for a transmission line. The trace is on the outside of the PCB and referenced to a ground plane as shown in Figure 6.40 with its characteristic impedance.

$$Z_0 = \frac{87}{\sqrt{\varepsilon_r + 1.41}} \ln\left(\frac{5.98h}{0.8W + 1}\right) \qquad (6.51)$$

TABLE 6.2 Trace thickness per PCB weight

PCB weight (oz/in.²)	Copper trace height (μm)
0.5	17.8
1.0	35.6
2.0	71.1
3.0	106.0

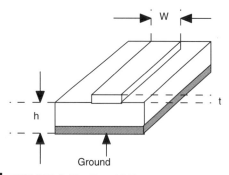

FIGURE 6.40 Typical PCB transmission line in the form of a microstrip.

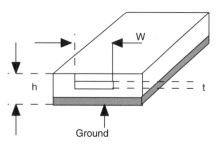

FIGURE 6.41 Embedded microstrip often used in PCB.

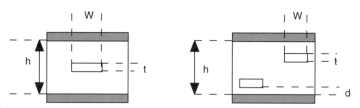

FIGURE 6.42 Stripline configuration usage in PCB traces.

The embedded microstrip is very similar to the microstrip but with a pads-only layer covering the signal conductor as shown in Figure 6.41.

$$Z_0 = \frac{K}{\sqrt{0.805\varepsilon_r + 2}} \ln\left(\frac{5.98h}{0.8W + 1}\right) \quad (6.52)$$

where $60 \leq K \leq 65$. The stripline features a trace sandwiched between two conductors, one above and one below, which act as ground plane. In the dual stripline, there are two levels of signal traces sandwiched between two ground planes as shown in Figure 6.42.

For Figure 6.42 with a single stripline,

$$Z_0 = \frac{60}{\varepsilon_r} \ln\left(\frac{4h}{0.67\pi\omega(0.8 + t/W)}\right) \quad (6.53)$$

THE EFFECT OF TRACE RESISTANCE

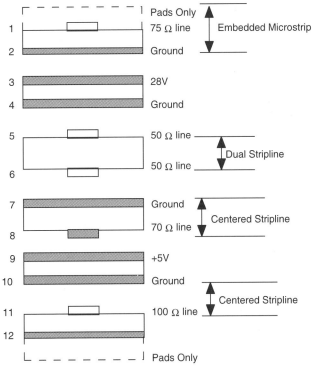

FIGURE 6.43 Cross-sectional view of a 12-layer PCB.

For Figure 6.42 with a double stripline,

$$Z_0 = \frac{2f_1 f_2}{f_1 + f_2}$$
$$f_1 = \frac{60}{\sqrt{\varepsilon_r}} \ln\left(\frac{8d}{0.67\pi\omega(0.8 + t/W)}\right) \quad (6.54)$$
$$f_2 = \frac{60}{\sqrt{\varepsilon_r}} \ln\left(\frac{8h}{0.67\pi\omega(0.8 + t/W)}\right)$$

Figure 6.43 shows an example of a 12-layer board containing some of the kinds of traces already discussed.

Comparing these types of transmission lines, we see the microstrip is the best solution for clock skew because signal travels faster on a microstrip than on a stripline (7 in./nS vs 5.5 in./nS), which is about 40% faster than the striplines. This is due to the fact that the stripline has twice the capacitance as the microstrip line and the propagation time per unit length is proportional to the square root of the product of inductance and capacitance per unit length. The quietest signal line is the one sandwiched between two planes, but such lines are not often used in PCB design because they require thicker boards. Therefore, it is better to use dual embedded striplines, and to avoid crosstalk, such lines should be as much perpendicular to each other as possible and certainly not next to or parallel to each other.

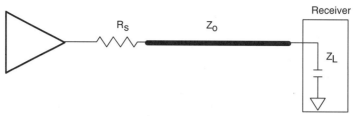

FIGURE 6.44 Impedance matching of transmission lines for driver–receivers.

Manipulation of trace widths for a given dielectric thickness in a PCB can be used to match impedances of IC logic since different logic families have different impedance requirements. Most designs, however, provide separate layers for the signal distribution of a given logic family and use another layer for another class of logic families. Furthermore, instead of changing trace width to match impedance, the thickness of the PCB layers is changed instead to accomplish the same objective without affecting the structural integrity of the PCB layout.

To put the finishing touches of impedance matching between the PCB trace and the load, a novel approach is to terminate the source end, as shown in Figure 6.44. This approach has an EMI advantage because it limits high-frequency periodic currents. A resistor is added to the transmission line between the driver and the line. The ideal value of the series resistor would match the impedance of the transmission line. As shown in the figure, most likely the incident wave on the load will be reflected back but once the reflection is received back at the source it is clamped down as the reflected wave is terminated on the series resistor. Once the current transition is complete, very small current continues. The main disadvantage of these series resistors is that if we must use ones that are large enough in value to match the trace impedance, we will generally limit the switching time of the line.

6.14. ASIC SIGNAL INTEGRITY ISSUES (GROUND BOUNCE)

CMOS semicustom ASICs often perform better than board designs in PCB with less power consumption. However, these power consumption advantages also produce some adverse effects. As larger CMOS drivers attempt to match bipolar driver speeds and current-carrying capabilities, the noise induced by simultaneous switching is becoming a problem for ASIC designers.

Let us consider Figure 6.45, where glitches in the TRBBUS bus and TRPBUS buses are experienced. It is estimated that such glitches are the result of grounding problems including ground bounce. Ground bounce is viewed in the scenario of driving a bus, simultaneously switching from high (5.0 V) to low (0.4 V). The inductive ground positive swing up drives n-channel transistors low to switch on and off, thus yielding false "output highs." Other anomalies in the ground bounce cause a bidirectional buffer output that is high to feedback into the device, changing the circuit state.

ASIC SIGNAL INTEGRITY ISSUES (GROUND BOUNCE)

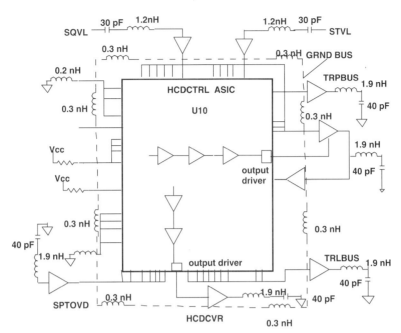

FIGURE 6.45 ASIC overview of its output losses and ground connections.

The equation given by

$$I_c(\text{switch}) = C_{\text{load}} \times \frac{dV_c(\text{switch})}{dt} \quad (6.55)$$

represents the current needed to switch a capacitive load from a given voltage level. The equation given by

$$V_{\text{tran}} = -L_{\text{pin}} \times \frac{dI_c(\text{switch})}{dt} \quad (6.56)$$

represents the transient voltage induced by the associated inductance of the switching current. This inductance is shown as pin inductance because of the contribution of the package pins to the overall inductance in the input–output circuits. If we combine these two equations, we obtain an expression for the transient-induced voltage:

$$V_{\text{tran}} = C_{\text{load}} \times L_{\text{pin}} \times \frac{d^2 V_c(\text{switch})}{dt^2} \quad (6.57)$$

This transient voltage is the noise input into the ground paths, which would couple to other pins and connectors. As we can observe, the noise voltage is dependent on the load capacitance, package inductance, and the switching speed. The load capacitance affects the voltage transition rate of the driver. The package inductance can include not only the pin inductance but all other associated parasitic inductances within the ASIC package. Notice that if the

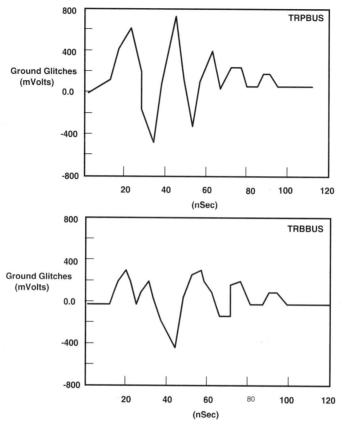

FIGURE 6.46 SPICE simulation results at ground pins of TRPBUS and TRBBUS output.

C_{load} increases, the voltage transition rate slows down and this may upset the overall V_{tran}, because $d^2V_c(switch)/dt^2$ decreases as L_{load} increases.

A SPICE analysis simulation shows TRPBUS and TRBBUS outputs from the ground pins as they experience the transient voltage, as seen in Figure 6.46.

In the preceding analysis, we have neglected impedance matching of the output drivers. Mismatched impedances cause reflections that can result in a fault system. Traces in low-frequency circuits are no more than conduits of currents. As switching speed increases, traces exhibit a new behavior, acting instead as transmission lines with defined characteristic impedance. To control the RF currents, we must therefore treat longer traces as transmission lines and design these accordingly.

Traces should be designed as transmission lines at high frequencies where the traces are long enough so that the reflection from the far end of the transmission line is delayed, arriving later than the original transition time. A trace can be considered as a transmission line if

$$T \leq 2L \times t_L \tag{6.58}$$

where T is the output transition time (rise or fall times), L is the length of PCB trace, and t_L is the loaded transmission delay/unit length.

In Figure 6.45, the system load capacitance of 40 pF is what the device would see in operation. The 1.9-nH values represent the lead wire inductance from the IC to the package used in the ASIC assembly. The 0.3- and 0.5-nH values are the inductance values of the ASIC peripheral internal ground bus. They will sink all driving current through the ground pins. The inductance is determined from the sum of the wire inductance (round wire assumed) which is about 1.27 nH/in. and the wire inductance over a ground plane

$$L_{\text{total}} = L_W + L_{Wg}$$
$$= 1.27\,\text{nH/in.} + (0.005)\ln\left(\frac{4h}{d}\right)\,\mu\text{H/in.} \quad (6.59)$$

where L_W is the internal round wire inductance, L_{Wg} is the wire inductance over a ground plane, h is the height above ground plane, and d is the wire diameter.

If the lead wires are 1.7 mils in diameter and 35 mils above the ground plane with a length of 80 mils the inductance is approximately given by 1.9 nH, which was the value used in the lead wire inductance. The ground bus of this ASIC is 180 μm wide and approximately 1.2 μm thick. We can calculate the equivalent diameter

$$D = 2(0.335)W\left(0.8 + \frac{t}{W}\right) \quad \text{for } 0.1 \le t/W \le 0.8 \quad (6.60)$$

6.15. CROSSTALK THROUGH PC CARD PINS

In digital systems, as previously stated, the electrical noise is caused by the coupling of electromagnetic energy from a source circuit into a victim or susceptible circuit. In a digital system, noise can cause the binary state of any given signal to shift involuntarily. These kinds of shifts can be transient (glitch) or may last for a longer time such as in a pulse. Transient noise, for example, may appear as a glitch that may show up as a clock transition in a flip-flop. One of the most important sources of noise is the crosstalk.

Crosstalk is defined as the coupling of voltage to an adjacent line through mutual coupling composed of either a mutual inductance or a coupling capacitance or both. The coupled voltage adds or subtracts to the actual signal voltage, thus moving a circuit closer to the switching threshold. Figure 6.47 illustrates the mutual inductance and capacitance phenomena.

Mutual inductance is created by a magnetic field generated by a current flowing through a current loop. Mutual inductance injects a noise voltage into an adjacent circuit proportional to its rate of change of current. The noise voltage can be represented by the equation

$$V_n = L_m \frac{dI(t)}{dt} \quad (6.61)$$

The coupling stray capacitance is rated by an electric field generated by a voltage change experienced in the noisy circuit. This time-varying field between

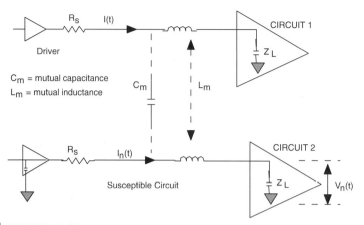

FIGURE 6.47 Representation of mutual inductance and capacitance between noisy and susceptible circuits.

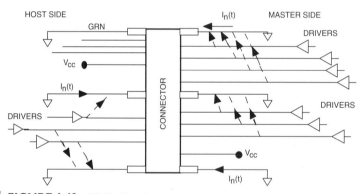

FIGURE 6.48 PC Card bus interconnect system.

the two conductors can be represented by a capacitor connected between the two conductors. This coupling capacitance injects a noise current into an adjacent circuit proportional to its rate of change of voltage.

$$I_n(t) = C_{12}\frac{dV_c(t)}{dt} \qquad (6.62)$$

There are three main noise sources in PC cards: (1) noise through crosstalk due to the mutual inductance and/or capacitance among the connector pins, (2) ground shifts in the connector as a result of the high number of switching outputs per ground pin on the connector, and (3) V_{cc} shifts due to the large number of switching outputs per V_{cc}. The ground shifts and the V_{cc} shifts occur as a result of the large number of outputs on the card switching simultaneously. Figure 6.48 shows the PC card bus interconnect system.

From Eq. (6.62),

$$I_n(t) = C_m\frac{dV_c(t)}{dt} \qquad (6.63)$$

and its first derivative, given by

$$\frac{dI_n(t)}{dt} = C_m \frac{d^2 V_c(t)}{dt^2} \tag{6.64}$$

The dV_c/dt in a circuit is related to the 10–90% rise time and the voltage swing ΔV.

$$\frac{dV_c(t)}{dt} = \frac{\Delta V}{T_{10-90\%}} \tag{6.65}$$

The maximum of $I_n(t)$ is given by its first derivative and obtained as

$$\text{maximum } I_n(t) = C_m \frac{1.52 \Delta V}{T_{10-90\%}^2} \tag{6.66}$$

Maximum inductive crosstalk is given by

$$\begin{aligned} V_n &= L_m \frac{dI(t)}{dt} = L_m C_m \frac{d^2 V_c(t)}{dt^2} \\ V_n &= \text{crosstalk} = \frac{L_m C (1.52 \Delta V)}{T_{10-90\%}^2} \end{aligned} \tag{6.67}$$

Let us consider, for example, the case of Figure 6.47, where the loop size of each circuit is 6 in. long by 0.5 in. high, running in parallel (maximum coupling) at a separation of 0.2 in., and the mutual inductance L_m is given by equation

$$L_m = L \left[\frac{1}{1 + (d/h)^2} \right] = (1.27 \text{nH/in.}) \left[\frac{1}{1 + (0.2/0.5)^2} \right] = 109.5 \text{ nH}$$

If $C = 30$ pF is the typical load capacitance, $\Delta V = 3.7$ V, and $T_{10-90\%} = 5.0$ nS (estimated). The crosstalk obtained from Eq. (6.67) is 0.73 V.

6.16. PARASITIC EXTRACTION AND VERIFICATION TOOLS FOR ASIC

At 0.5 μm, the comparison of what is designed with present EDA tools and what is delivered is not a real challenge, even as the frequency of the clocks increases. In 0.5-μm design rules, these obstacle are manageable. Many existing tools are beginning to show signs of dealing with 0.35-μm design, but the 0.25-μm barrier is still (as of this writing) out of reach for EDA tools. The deep submicrometer area, as the 0.25-μm region is known, shows special difficulties for designers. The main problem is that concepts such as capacitances and resistance, which were not crucial in the past, have now become essential to the success of a good design.

Parasitic extraction is part of the design process where parasitic effects such as capacitance, inductance, and resistance are taken into account. At 0.5 μm and above, gate delays dominate over the delay that could be experienced by the interconnects. The work on parasitic extraction may be only on a small portion of a circuit and it is only an estimate. The device resistive and capacitive effects can be viewed in terms of a lumped RC model. These approximations are acceptable because the overall effects on the parasitics are small and any error between the lumped RC model estimates and the real parasitic effects is small, causing no adverse effects in the design. As we move to 0.35 μm, the problems begin to arise, and by the time we get to 0.25 μm, the current methodologies are of very little use. At this point, the interconnect delays dominate over gate delays and the physics of the interactions between components such as metal layers, wires, and transistors play a major role. Therefore, in the design process, we can no longer use RC-lumped parameter models. Interconnects could affect a circuit's behavior and can easily account for a great percentage of the overall performance. Designers must better assess the values of these parasitics for paper design. The two most important factors in these processes become (1) good approximation of parasitics and (2) a methodology for analyzing the large amount of data.

Tweaking present extraction tools is not as easy as it sounds because parasitic extraction in deep submicrometer design is very complicated. For example, metal cross sections become an issue because the width and height are equal at 0.35 μm and at 0.25 μm. The aspect ratio of height to width is 1:2 and this trend will continue, and the lumped-c model no longer works. Determining interconnect parasitics in a deep submicrometer design requires two steps. The first step is process technology characterization as models or libraries, to build estimates for the possible parasitics. The second step is full chip extraction, which would determine the location of the parasitics, and generate the libraries for network analysis.

Interconnect characterization involves using a variety of field solvers to determine the magnitude and location of capacitance and resistance. These field solvers discretize the layout structure in three dimensions and solve Laplace

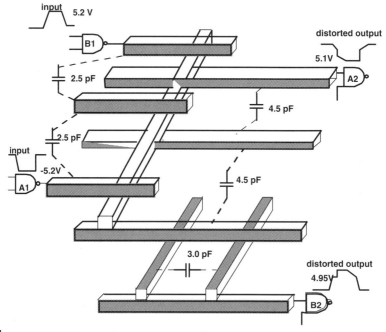

FIGURE 6.49 Representation of deep submicrometer parasitics in ASIC.

equations using finite element methods, the finite difference time domain, the boundary element method, Monte Carlo methods, moment methods, and fast multipole methods. Unfortunately, these kinds of analyses require large amounts of data and enormous amounts of computer time. Therefore, we have quasi 2D and 3D views of interconnects. The quasi 3D profiles are stored in libraries with their capacitances. In the extraction process, the patterns of interconnects are matched to the library patterns to calculate parasitics. For example, the nets can be broken and divided into 3D sections, generating a representation library that totals the whole structure in 3D for each process. Closed-form look-up tables can be used to match the values in libraries, providing extraction times similar to 2D-based solutions. An example of deep submicrometer parasitics and their effects in design is shown in Figure 6.49.

Notice from the figure that the output of $A2$ is corrupted as a result of parasitic capacitances among rails 3, 7, and 8 and between rails 1 and 4. The metal-to-metal capacitances are large enough that many of the interconnects share signals with their neighbors, resulting in irregular signals like the ones at the outputs of $A2$ and $B2$.

REFERENCES

Analog Devices. 1995. *Linear Design Seminar*. Prentice Hall, Englewood Cliffs, NJ.
Graeme, Jerald, and Bonnie Baker. 1996. "Design equations help optimize supply bypassing for op-amps." *Electronic Design*, June.

7
HARDWARE APPROACH TO DIGITAL SIGNAL PROCESSING

7.1. DISCRETE FAST FOURIER TRANSFORM

The discrete Fourier transform (DFT) is a mathematical operation that is performed on a finite length of contiguous discrete time samples to produce an equivalent number of frequency samples. Signal levels, noise levels, and harmonic content can all then be calculated from the DFT output. The fast Fourier transform (FFT) is simply an algorithm that is used to greatly reduce the number of mathematical calculations needed to perform the DFT output spectrum.

To calculate the DFT, a spectrally pure sine wave is applied to an ADC, and a number of contiguous samples are stored in a buffer memory. The record time contains an integer number of cycles of the sine wave, and time-weighting of the samples is required to reduce frequency side lobes. Without weighting, the discontinuity produced by not having an integer number of cycles will cause the main lobe energy to "leak" into many other frequency bins, making accurate spectral measurement impossible. A popular weighting function is called the "Hamming" function and is given by

$$W_n D_n + D_n \left[0.5 - 0.5 \cos\left(\frac{2\pi n}{M}\right) \right] \quad (7.1)$$

where $W_n D_n$ is the nth weighted data sample, D_n is the nth input data sample, and M is the total number of samples. In the usage of this weighting function,

*Thanks are expressed to Analog Devices for the usage of some of their components publications.

the leakage energy can be compressed into a small band of frequencies centered on the fundamental sine wave frequency.

Second, the program must find the DFT of the sequence of weighted data samples for $M/2$ frequencies. To do that, the program must solve the following two equations for the Kth frequency:

$$A_K = \frac{1}{M} \sum_{n=1}^{M} W_n D_n \cos\left[\frac{2\pi K(n-1)}{M}\right] \quad B_K = \frac{1}{M} \sum_{n=1}^{M} W_n D_n \sin\left[\frac{2\pi K(n-1)}{M}\right] \tag{7.2}$$

where A_K and B_K represent the magnitude of the frequency domain representation of the M time samples. The resolution or spacing between the spectral lines is given by the equation

$$\text{magnitude}_K = \sqrt{A_K^2 + B_K^2} \tag{7.3}$$

The results yield $M/2$ components, which are the frequency-domain representation of the M time samples. The resolution or spacing between the spectral lines is given by the equation

$$\Delta f = \frac{f_s}{M} \tag{7.4}$$

where M is the total record length. The value Δf is often referred to as the "bin" size.

The overall ADC SNR is then calculated by

$$\text{SNR} = 20 \log \left[\frac{\text{rms signal level}}{\text{rms noise level}}\right] \tag{7.5}$$

It is often useful to measure the third-order intermodulation products for two sine waves of frequencies f_1 and f_2, which are

$$\text{harmonic distortion} = 20 \log \left[\frac{\text{rms signal level}}{\text{rms harmonic level}}\right] \tag{7.6}$$

Total harmonic distortion is often calculated by root-sum squaring the first five harmonics of the fundamental sine wave and using the resulting number in the preceding formula for the rms harmonic level. These products occur at frequencies (see Figure 7.1)

$$\begin{array}{ll} 2f_1 + f_2 & 2f_2 + f_1 \\ 2f_1 - f_2 & 2f_2 - f_1 \end{array} \tag{7.7}$$

Most IMDs can be filtered out. However, if the two tones are of similar frequencies, the third-order IMD ($2f_1 - f_2, 2f_2 - f_1$) will be very close to the fundamental frequencies and cannot be easily filtered. The level of these products is of most concern in narrow-bandwidth applications.

7.2. DETERMINING THE PROPER FFT RECORD LENGTH

The first consideration in choosing the number of time-domain samples required or record length N is the required spectral resolution, Δf. Also, to perform the

DETERMINING THE PROPER FFT RECORD LENGTH

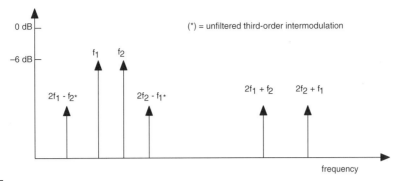

FIGURE 7.1 Third-order intermodulation products.

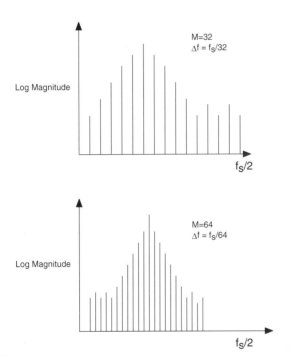

FIGURE 7.2 FFT spectrum vs f_s.

FFT computations, N must be equal to an integer that is a power of 2. The FFT spectral resolution Δf is given by

$$\Delta f = \frac{f_s}{M} \qquad (7.8)$$

where f_s is the sampling rate. Typically, M is selected to be between 256 and 4096, depending on the desired resolution and the amount of buffer memory available.

If windowing is used, the main lobe width and side lobe rolloff (leakage) are measured in units of "bin width" f_s/M, thereby leaving a larger percentage of the Nyquist spectrum uncontaminated, as shown in Figure 7.2.

Another consideration in determining the proper record length is the FFT noise floor itself. Assuming the digital signal processing (DSP) noise contribution is negligible, the rms signal to rms noise level in a single frequency bin of width Δf is given by the expression

$$\text{SNR}_{\text{FFT}} = 6.02N + 1.76\,\text{dB} + 10\log_{10}\left(\frac{M}{2}\right) \qquad (7.9)$$

7.3. COHERENT AND NONCOHERENT SAMPLING

The usage of coherent sampling eliminates leakage and the requirement for windowing. There are some requirements on the choice of sampling rate and the sine wave frequency:

$$\frac{f_{\text{in}}}{f_s} = \frac{M_c}{M} \qquad (7.10)$$

where

M_c = number of integer cycles of the sine wave during the record period
M = number of samples in record period
f_{in} = input sine wave frequency
f_s = sampling rate

For a whole number of cycles, M_c must be an integer. For nonrepetitive data M_c should also be odd and prime; that is, 1, 3, 5, 7, 11, 13, 17, etc. This ensures that all samples during the record period will be unique. When using coherent sampling, the ratio must be constant and this means that f_s and f_{in} are derived from sources that are tied to each other.

If noncoherent sampling is chosen, the input frequency selection has fewer restrictions. Integer submultiples of the sampling frequency should be avoided to prevent masking out harmonics of the fundamental. It is desirable to make the input frequency an odd multiple of the FFT frequency bin size.

The selection of the weighting function is to establish a trade-off between main lobe spreading and side lobe rolloff. The effects of windowing a sine wave using the Hamming window are shown in Figure 7.3.

Table 7.1 is useful for determining the number of samples to be included in calculating the energy in the fundamental. For a 12-bit ADC with a theoretical SNR of 74 dB, it would be appropriate to include 20 samples on either side of the fundamental in calculating the rms signal energy (if Hamming weighting is used).

To make efficient use of the FFT processor, the weighting function coefficient should be calculated one time and then stored in a lookup table. If better resolution is needed, the minimum four-term Blackman–Harris function is required, as given by the expression

$$a_n = a_0 - a_1\cos\left(\frac{2\pi n}{M}\right) + a_2\cos\left(\frac{2\pi\cdot 2n}{M}\right) + a_3\cos\left(\frac{2\pi\cdot 3n}{M}\right) \qquad (7.11)$$

COHERENT VS NONCOHERENT SAMPLING

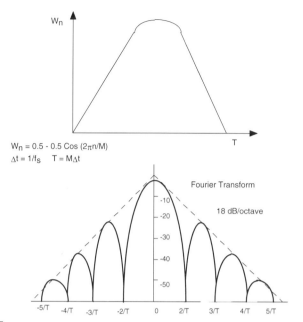

FIGURE 7.3 Time and frequency illustration of the Hamming window.

TABLE 7.1 Number of samples needed for calculating the fundamental

Bins from fundamentals	Side lobe attenuation (dB)
2.5	32
5.0	50
10.0	68
20.0	86

where $a_0 = 0.35875$, $a_1 = 0.48829$, $a_2 = 0.14128$, and $a_3 = 0.01168$. Figure 7.4 shows a spectral plot of this function.

7.4. COHERENT VS NONCOHERENT SAMPLING

The primary application of coherent sampling using FFT is in the testing ADCs using sine wave inputs. If the proper ratios between f_{in} and f_s are observed, the need for windowing is eliminated, as previously discussed. This greatly increases the spectral resolution of FFT and creates an ideal environment for critically evaluating the spectral response of the ADC. Great effort must be made, however, to ensure the spectral purity and phase stability of f_{in} and f_s. The choice of the frequencies and their proper ratio is somewhat tedious and high-quality frequency synthesizers are required to generate the phase-locked signals (Figure 7.5).

190 7 HARDWARE APPROACH TO DIGITAL SIGNAL PROCESSING

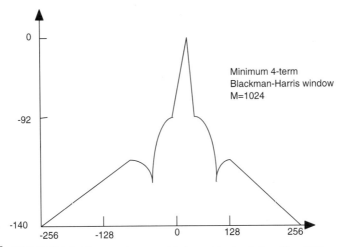

FIGURE 7.4 Minimum four-term Blackman–Harris window, $M = 1024$.

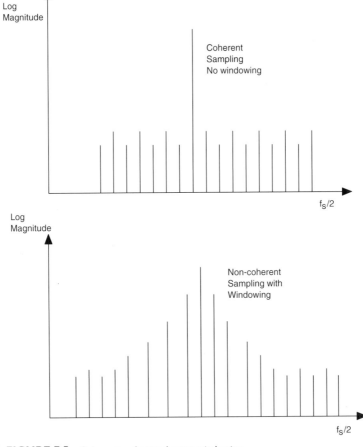

FIGURE 7.5 Coherent and noncoherent windowing.

In practical ADC applications, the precise frequency content of the signal being digitized is not usually known. If the ADC is being used in a spectral analysis application, a windowing function is required to process the signal.

Coherent sampling requires careful attention in the selection of the frequencies, while noncoherent sampling requires careful attention and use of the windowing function.

7.5. ULTRASOUND APPLICATION

The block diagram of a typical ultrasound system is shown in Figure 7.6. A burst ultrasound energy (1 to 13 MHz) is generated in an electromechanical piezoelectric transducer, which is in contact with the surface of the body. The velocity of propagation of ultrasound waves in most soft body tissues is around 1500 m/s. Returning echoes are produced at the interfaces between the various types of soft-body tissues. The round-trip time of each echo is used to determine its distance from the transducer.

The soft-body tissues in the body can alternate the burst of ultrasound by about $1\,dB/cm\,MHz^{-1}$. For thicker parts of the body as in the abdominal region, frequencies of 1–2 MHz are often used. For imaging of shorter path lengths, frequencies as high as 20 MHz can be used.

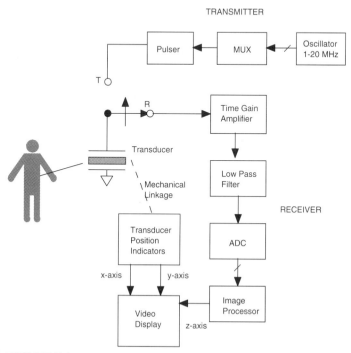

FIGURE 7.6 Ultrasound system block diagram.

Due to the large attenuation involved, the receiver transducer must have a dynamic range of 150 dB when scanning from 1 to 10 cm deep at 10 MHz, independent of tissue variations that may be encountered. For this reason the output of the transducer is applied to a time-gating amplifier whose gain in decibels is directly proportional to the amount of time elapsed from the transmission of the burst, as shown in Figure 7.7.

Near the surface, there is very little attenuation. For deep signal returns, gain of the signal is required in the receiving end to compensate for the path attenuation. The time gain amplifier compensates for normal signal attenuation related to the delay distance. The receiver ADC "sees" only the intensity variations associated with different types of tissues. For body tissues that are soft throughout (i.e., constant attenuation), a fixed gain versus time is most adequate. In other situations involving blood pools or fixed regions, it is desirable to have variable gain and the machine operator can usually make these changes.

In phase array ultrasound systems, the angular information is determined by phasing delays from a number of transducers (transmitting and receiving) to electronically select the angle to be processed. The delays at the transmitter and receiver are adjusted using variable delay filters. Low-cost, low-power, high-performance analog-to-digital converters and DSP technology can make it practical to digitize RF directly and digitally control the delay requirements. This technique, known as digital beam forming, is shown in Figure 7.8 and represents the future of ultrasound.

The Analog Devices AD600 is a low-noise variable-gain amplifier for use in ultrasound systems. A block diagram is shown in Figure 7.9. The AD600 can be configured as a dual channel device providing 80 dB of gain. Each variable-gain amplifier contains a low-distortion fixed-gain (41-dB) feedback amplifier, preceded by a voltage-controlled attenuator (0 to 42 dB) and gain control circuitry

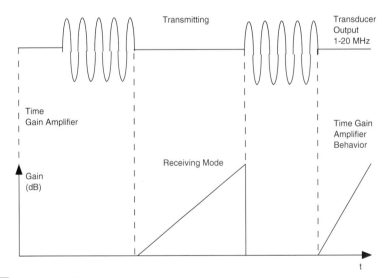

FIGURE 7.7 Transmission burst in ultrasound systems.

ULTRASOUND APPLICATION

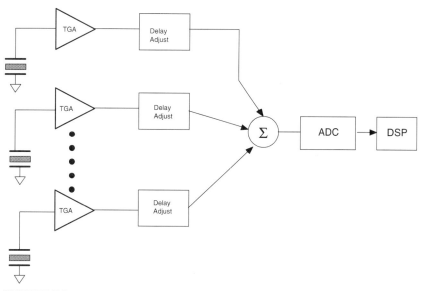

FIGURE 7.8 Digitizing and processing RF signal directly.

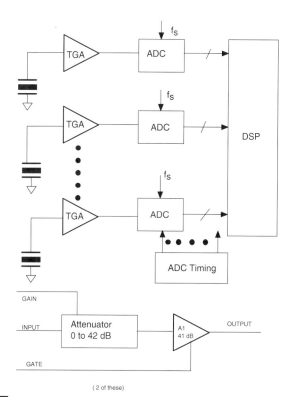

FIGURE 7.9 Low-noise-gain amplifier for ultrasound use.

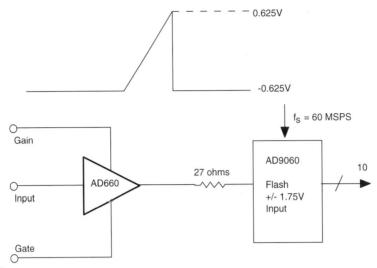
FIGURE 7.10 AD600–AD9060 interface for ultrasound.

for smoothly interpolating the attenuator. The differential high-impedance control inputs have a scale factor of 32 dB/V. Each amplifier has an independent gating function that blocks transmission and sets the amplifier's dc output level to within a few millivolts of the ground output. An internal voltage reference is used to calibrate all the internal parameters. The 3-dB bandwidth of each variable amplifier is nominally 35 MHz and independent of the gain setting.

In Figure 7.10 we observe the AD600 interface with the AD9060 (10 bits, 60 MSPS) A–D flash converter. The gain control input is linear ramp, which is generated at the appropriate time with respect to the transmission. To provide the same dynamic range without the thermo gravimetric analysis (TGA) (i.e., the ADC interfaces directly to the transducer via a fixed-gain amplifier), the ADC would require a dynamic range of about 100 dB. This implies a 16-bit ADC, which would need to operate at least at 30 MSPS.

7.6. DISCRETE TIME SAMPLING OF ANALOG SIGNALS

The sampling of an analog signal is illustrated in Figure 7.11. The continuous analog data must be sampled at discrete intervals t_s, which must be chosen carefully to demonstrate an accurate representation of the analog signal, as shown in Figure 7.11. The sampling rate is dictated by the Nyquist criteria.

1. An analog signal with a bandwidth of BW_a must be sampled at a rate of $f_s > 2BW_a(f)$ to avoid the loss of the information.
2. If $f_s < 2BW_a(f)$ a phenomenon called aliasing will occur.

The effects of aliasing on the dynamic range of a sampled data system are shown in Figure 7.12. The top part of the figure shows the desired condition at

DISCRETE TIME SAMPLING OF ANALOG SIGNALS

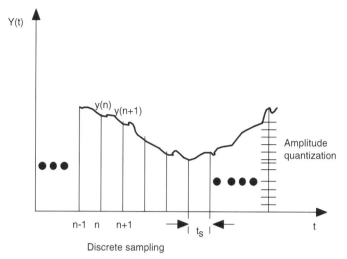

FIGURE 7.11 Example of sampling an analog signal.

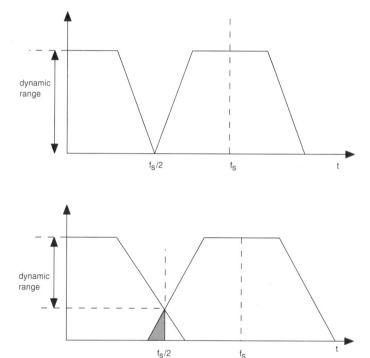

FIGURE 7.12 Example of aliasing in sample systems.

the Nyquist point, where below is the desired dynamic range. The lower part of the figure shows the effect of aliasing, where the upper frequency dynamic range is limited by the aliased components.

When we have aliasing, there is a reduction in the overall signal-to-noise ratio at the higher frequencies, which could result in the distortion due to

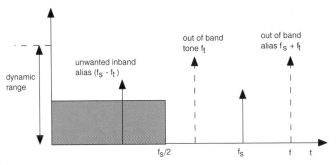

FIGURE 7.13 Out-of-band tones of harmonics.

aliased out-of-band tones of harmonics as shown in Figure 7.13. Simply stated, dynamic range is the ratio of the largest expected signal to the smallest signal that must be resolved and it is usually expressed in decibels.

Antialiasing filters are an important consideration for improving signal-to-noise ratio. The following are some rules of thumb for specifying the filter:

1. Set the corner frequency of the antialiasing filter equal to the desired analog input bandwidth, $BW_a(f)$. This means that the passband of the filter $f_{\text{pass}} = BW_a(f)$.
2. Define the beginning of the filter's stop band $f_{\text{stop}} = f_s/2$.
3. Let the filter stop band attenuation be the desired upper frequency dynamic range. These parameters define the transition band characteristics of the filter; that is, it must achieve a stop band attenuation equal to the dynamic range over $\log_2(f_{\text{stop}}/f_{\text{pass}})$ octaves.
4. The appropriate order of the filter M (the number of poles) required to achieve this transition band slope can be determined by the expression

$$M = \frac{\text{dynamic range}}{6\log_2[f_s/(2BW_a(f))]} \tag{7.12}$$

7.7. DIGITAL SIGNAL PROCESSING TECHNIQUES

One of the fundamental processes in DSP is the development of digital filters. Analog filters are limited in their performance and are susceptible to passive component fluctuations over time and temperature. The characteristics of digital filters can be changed by software control. This means that digital filters have a great advantage in the processing of signal from sensors, digital audio, mobile radio, etc.

The procedure for designing digital filters is very similar, conceptually, to the design of analog filters. The desired filter responses are characterized by the transfer function and phase response. The main difference between analog and digital filters is that instead of calculating resistor, capacitor, and inductor values for an analog filter, coefficient values are calculated for a digital filter. Thus, for a digital filter, numbers replace the physical resistor and capacitor components of the analog filter. These numbers will reside in memory as filter

FINITE IMPULSE RESPONSE DIGITAL FILTERS

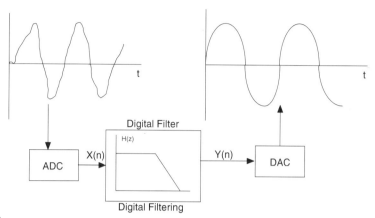

FIGURE 7.14 Illustrating digital filtering.

coefficients and are used along with data values from the ADC in performing the filtering process. Because the digital filter is designed as a discrete function, it works only with digitized data for each sampling period. Figure 7.14 illustrates the basic filtering function. It shows a low-frequency signal containing high-frequency noise, which must be filtered out. The wave force representing the signals is first digitized with an ADC to produce samples $x(n)$. The data values are fed to a digital filter, which in this case is a low-pass filter. The output data samples $y(n)$ are used to reconstruct the analog waveform using a DAC.

To maintain real-time operation, the DSP processor must be able to execute all the steps in the filter routine within one clock period $1/f_s$. Sometimes, this is very difficult to do and the high-speed ADC data are stored in a buffer memory. The buffer memory is then read at a rate that is compatible with the speed of the DSP-based digital filter.

7.8. FINITE IMPULSE RESPONSE DIGITAL FILTERS

The simplest digital filter is the finite impulse response (FIR) filter, and the most elementary form of an FIR filter is a moving-average filter as shown in Figure 7.15. In the figure we show a nine-sample moving average of a function plotted along the k axis. After nine samples are collected, the first data point on the moving average is computed by adding the nine data samples together and dividing by 9. Another way to view the process is to weight each sample by a factor of 1/9 and perform a summation. To obtain the second point in the moving average, the first weighted data sample is subtracted from the summation, and the tenth weighted sample is added to the summation. This process continues and can be viewed as a very rough low-pass filtering. The digital implementation of the process is presented in Figure 7.16, which shows the various multiplications and delays and the summation.

When processing an electrical signal, a moving average might look like Figure 7.17. It is useful from a mathematical standpoint to view the moving-average filter as a convolution of the filter impulse response $h(t)$ with the

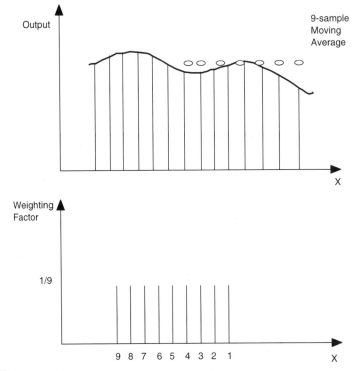

FIGURE 7.15 Simple moving average FIR filter.

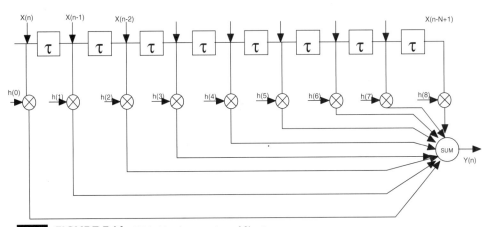

FIGURE 7.16 Digital implementation of filtering.

sampled data points $x(t)$ to obtain the output $y(t)$, as shown in Figure 7.17. For a linear convolution, the operation involves multiplying $x(t)$ by a reversed and linearly shifted version of $h(t)$, and then summing the values in the product.

The $\sin(x)/x$ frequency response of the moving-average filter is shown in Figure 7.18 for various numbers of taps, N, where N represents the number of sample points. Increasing the number of sample points sharpens the rolloff

FINITE IMPULSE RESPONSE DIGITAL FILTERS

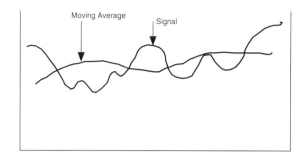

Moving Average FIR Filter Applied to Analog Signal

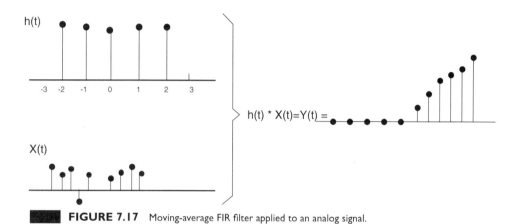

FIGURE 7.17 Moving-average FIR filter applied to an analog signal.

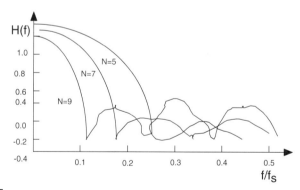

FIGURE 7.18 Frequency response of a moving-average filter for different number of samples.

characteristic of the moving-average filter but it does not improve the need for diminishing side lobes.

The most important issues in FIR filter design are the appropriate selection of the filter coefficients and the number of samples to realize the desired transfer function $H(f)$. Filter design software will be able to translate the frequency response $H(f)$ into a set of FIR coefficients. The coefficients of the FIR filter

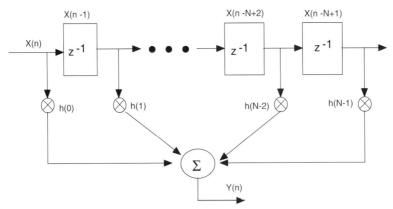

FIGURE 7.19 Illustration of an FIR filter.

$h(n)$ are the quantized values of the impulse response of the frequency transfer function.

The FIR filter (Figure 7.19) must perform the following convolution equation:

$$y(n) = h(n) * x(n) = \sum_{i=0}^{N-1} h(i)x(n-i) \tag{7.13}$$

where $h(i)$ is the filter coefficient array and $x(n-i)$ is the input data array to the filter. The number N is the number of samples of the filter. In the series FIR filter equations, the N coefficient locations are always accessed sequentially from $h(0)$ to $h(N-1)$. The associated data points circulate through the memory, adding new samples and replacing old ones.

In practice, the concepts presented above have been implemented in easy-to-use CAD programs that can run on most PCs. In these CAD programs, it's only necessary to specify the desired FIR filter characteristics (sampling frequency, passband frequency, passband ripple, and stopband attenuation), as shown in Figure 7.20. The CAD program calculates the number of filter samples required (N), the impulse response, and the filter coefficient. The number of samples must be compatible with the throughput of the DSP processor and the sampling rate.

7.9. INFINITE IMPULSE RESPONSE DIGITAL FILTERS

Digital FIR filters have no analog counterpart. In addition, FIR filters have only zeros and no poles. Infinite impulse response (IIR) filters do have traditional analog counterparts (Butterworth, Chebyshev, and elliptic) and can be analyzed and synthesized using more familiar traditional filter design techniques. Figure 7.21 shows a second-order low-pass active filter and its IIR digital filter equivalent is also shown in Figure 7.21. This second-order IIR filter is referred to as the biquad and forms the basic building block for most higher-order IIR designs. The difference equation which describes the characteristics of the filter

INFINITE IMPULSE RESPONSE DIGITAL FILTERS

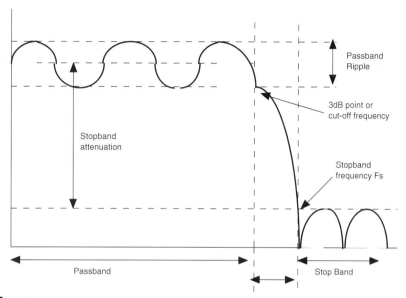

FIGURE 7.20 Characteristics of an FIR filter.

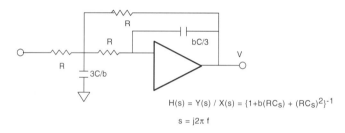

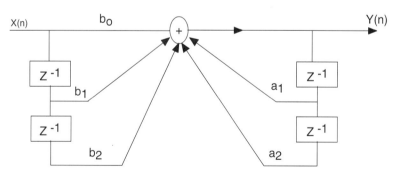

Y(n) = b₀ X(n) + b₁ X(n-1) + b₂ X(n-2) - a₁ Y(n-1) - a₂ Y(n-2)

FIGURE 7.21 The IIR filter.

with five coefficients is shown by equation (7.13a)

$$Y(n) = b_0 X(n) + b_1 X(n-1) + b_2 X(n-2) - a_1 Y(n-1) - a_2 Y(n-2)$$

$$H(s) = \frac{Y(s)}{X(s)} = \frac{1}{1 + b(RC_s) + (RC_s)^2} \quad s = j2\pi f \quad (7.13a)$$

The general digital filter equation shown in Figure 7.21 gives rise to the general transfer function $H(z)$, which contains polynomials in both the numerator and the denominator (see Equation 7.14). The roots of the numerator determine the zero locations. Although it is possible to construct a high-order IIR filter directly from this equation, accumulation errors due to quantization errors (finite word length arithmetic) may give rise to instability and large errors. For this reason, it is common to cascade several biquad sections with appropriate coefficients rather than use the direct form implementation. The biquads can be cascaded to minimize the coefficient of quantization and the recursive accumulation errors. Cascaded biquads are more stable and minimize the effects of errors due to finite arithmetic errors.

$$Y(n) = \sum_{K=0}^{M} b_K X(n-K) + \sum_{K=1}^{N} a_K Y(n-K)$$

$$H(z) = \frac{\sum_{K=0}^{M} b_K z^{-K} \quad \text{(zeros)}}{1 - \sum_{K=1}^{N} a_K z^{-K} \quad \text{(poles)}}$$

(7.14)

7.10. FAST FOURIER TRANSFORM

In many applications, it is necessary to process signals in the frequency domain. In the analog domain, this can be easily accomplished using an analog spectrum analyzer. Mathematically, this can be accomplished by obtaining the Fourier transform of the analog signal. The Fourier transform yields the spectral content of the analog signal.

In digital systems, however, this process needs to be accomplished by DSP processing of the ADC output data. Furthermore, there are two basic differences between the analog and digital spectral analysis. First, the output of the ADC consists of discrete, quantized examples of the continuous input $x(t)$. In sampled data systems, the discrete Fourier transform performs the transformation of the time domain signal into the frequency domain. Second, the DFT operates on a finite number of sampled data points, while the Fourier transform operates on a continuous waveform.

If $x(n)$ is the sequence of N input data samples, then the DFT produces a sequence of N samples $X(K)$ spaced equally in frequency. The DFT can be viewed as a correlation between the input signal to many sinusoids, evaluating the frequency content of the input signal.

$$X(K) = \sum_{n=v}^{N-1} x(n) e^{-j2\pi nk/N} \tag{7.15}$$

where $e^{-j2\pi nK/N} = \cos(2\pi nK/N) - j\sin(2\pi nK/N)$.

For an N-point DFT it would require N samples of the input signal and N points from a sinusoid form $-f_s/2$ to $+f_s/2$ are used. Each pass of the DFT checks the sinusoid against the input signal to see how much of that frequency is present in the input signal. This is repeated for each of the N frequencies (a large task if N is large). If the sampling frequency is f_s, then the spacing between the spectral lines is f_s/N, or $1/Nt_s$, where t_s is the sampling period, $1/f_s$. Spectral analysis is often performed with complex signals (i.e., having both real and imaginary components) so that phase information as well as amplitude and frequency information is obtained. The total number of multiplications required is N^2 (N complex data values multiplied with the N sinusoid values). This amount of computation is required when all output frequencies are to be calculated.

In most spectral analyses, the entire frequency spectrum up to $f_s/2$ must be computed. The FFT is nothing more than a fast algorithm to speed up the DFT calculations by reducing the number of multiplications and additions required. The FFT is based on taking advantage of certain algebraic and trigonometric symmetries in the DFT computational process. For example, if an N-point DFT is performed, N^2 complex multiplications are required. It is possible to split the N-point DFT into two $N/2$ points DFT and end up with same result, a process called decimation. Each $(N/2)$ DFT requires $(N/2)^2$ complex multiplications for a total of $N^2/4$ complex multiplications, which is a significant reduction.

Figure 7.22 shows an N-point DFT broken up into two $N/2$-point DFTs. The presence of a phase factor W on a horizontal line indicates a multiplication by W (W is often called the twiddle factor). The points where the arrows intersect the horizontal lines indicated a summation. The presence of a -1 on the line indicates a sign reversal.

The decimation process can continue; for example, each $N/2$-point DFT can be broken up into two $N/4$-point DFTs for an even greater reduction in computation. The decimation process can continue until the original DFT is broken up into two-point DFTs (the smallest DFT possible).

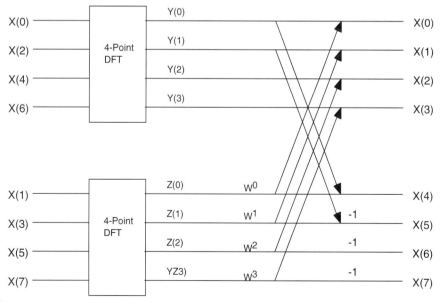

FIGURE 7.22 DFT implementation diagram.

204 7 HARDWARE APPROACH TO DIGITAL SIGNAL PROCESSING

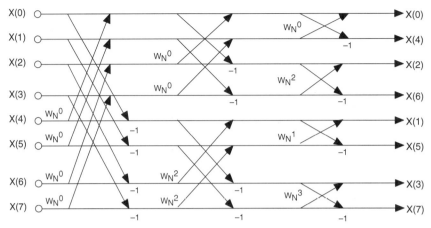

FIGURE 7.23 Eight-point decimation in time FFT normal order inputs, bit reversed outputs.

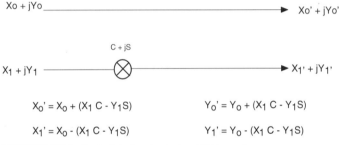

FIGURE 7.24 Radix-2 decimation in time FFT butterfly.

The final series of computations, after the decimation process is complete, is the FFT. This is shown for the eight-point DFT in Figure 7.23. Since the FFT was first decimated by a factor of 2, the FFT is known as a radix-2 FFT. If the initial DFT was decimated by a factor of 4, it would be referred to as a radix-4 FFT. Note that the input data points are taken in normal order, but the outputs are in bit-reversed order. Bit-reversed hardware is common in DSP processors such as the Analog Devices ADSP-2101. The basic calculation, essentially the two-point DFT, is commonly referred to as a butterfly calculation. The FFT is made up of many butterfly calculations. Figure 7.24 shows the basic butterfly for the radix-2 decimation-in-time FFT, which requires one complex multiply operation per butterfly.

7.11. FFT HARDWARE IMPLEMENTATION

In general the memory requirements for an N-point FFT are N locations for real data, N locations for imaginary data, are N locations for the sinusoidal data (also known as the FFT coefficients). As long as the memory requirements are met, the DSP processor must perform the needed calculations in the required

TABLE 7.2 Maximum sampling rate vs FFT execution time

FFT	Execution time (ms)	Maximum sampling rate (kHz)
256	0.59	434
512	1.3	394
1024	2.9	353
2048	6.5	315
4096	14.2	288

time. When comparing FFT specifications, it is important to make sure that the same type of FFT is used in all cases.

The first step in designing an FFT is to determine the number of points required N, or the record length. The sampling rate f_s must be at least twice the maximum input signal frequency of interest. The spectral resolution of the FFT is then given by f_s/N. The more points in the FFT, the better the spectral resolution. The signal must be divided up into windows T_w, short enough to ensure that individual features are not averaged out in the FFT, but T_w must be long enough to given adequate spectral resolution. It has been determined, for example, that for human speech, 20 ms is adequate, hence $T_w = 20$ ms. The number of sample points in the window T_w is equal to $T_w f_s = 20$ ms $\times$ 8 kHz $= 160$ points, which is rounded up to 256 points. This means that the DSP processor must complete the 256-point T_w. Otherwise real-time processing is not possible, and the computation would have to be done off line. Table 7.2 shows the maximum sampling rates for real-time operation associated with the FFT execution times.

7.11.1. DSP Hardware

In this section we study the architecture of the Analog Devices ADSP-2101. Many other DSP architectures follow similar design blocks.

An overall block diagram of the ADSP-2101 is shown in Figure 7.25. The ADSP processor contains three independent units: the arithmetic logic unit (ALU), the multiplier–accumulator (MAC), and the shifter. The computational unit processes 16-bit data directly and has provisions to support multiprecision computations. The ALU performs a standard set of arithmetic and logic operations. The MAC performs single-cycle multiply, multiply–add, and multiply–subtract operations. The shifter performs logical and arithmetic shifts, normalization, denormalization, and derive exponent operations. The shifter can be used to implement numeric format control including multiword floating-point representations.

The internal result (R) bus directly connects the computational units such that the output of any unit maybe the input of any unit in the next cycle. The program sequence and dedicated data address generators ensure efficient use of these computational units. The data address generators handle address pointer

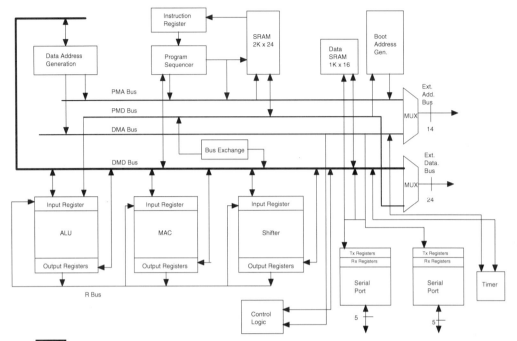

FIGURE 7.25 Block diagram of the AD2101 DSP from Analog Devices.

updates. Each data address generator keeps track of four address pointers. When the pointer is used to access data (indirect addressing), it is postmodified by the value of a specified modify register. A length value is associated with each pointer.

A length value may be associated with each pointer to implement automatic modulo addressing for circular buffers. The processor can generate two addresses simultaneously for dual operand fetches. The circular buffering feature is also used by the serial parts for automatic data transfers.

Efficient data transfer is achieved with the use of five internal buses. The two address buses (PMA and DMA) share a single external address bus, and the two data buses (PMD and DMD) share a single external data bus. Program memory can store both instructions and data, permitting the ADSP-2101 to fetch two operands in a single cycle, one from program memory and one from data memory as well as an instruction from program memory. Because the on-board program memory is so fast, the ADSP-2101 can fetch an operand from program memory and perform the next instruction in the same cycle.

The memory interface supports slow memories and memory-mapped peripherals with programmable wait state generation. One execution mode allows the ADSP-2101 to continue running while the buses are granted to another master as long as an external operation is not required. The other execution mode requires the processor to halt while the buses are granted.

The two serial ports provide a complete serial interface with companding in hardware and a wide variety of framed and frameless data transmit and data transmit and receive modes of operation. Each port can generate an internal

FFT HARDWARE IMPLEMENTATION

programmable serial clock or accept an external serial clock. Boot circuiting provides for loading on-chip program memory automatically from byte-wide external memory.

7.11.2. Arithmetic Logic Unit

The ALU is shown in Figure 7.26. The ALU provides a standard set of arithmetic and logic functions: add, subtract, negate, increment, decrement, absolute value, AND, OR, exclusive OR, and NOT. Two divide primitives are also

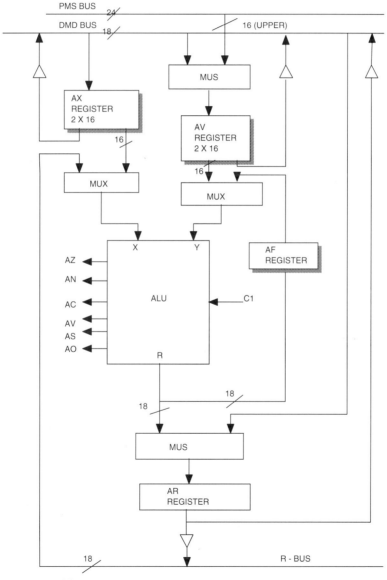

FIGURE 7.26 The ALU block diagram for the AD2101.

provided. The ALU takes two 16-bit inputs, X and Y, and generates one 16-bit output, R. The carry-in feature enables multiword computations. Six arithmetic status bits are generated: AZ (zero), AN (negative), AV (overflow), AC (array), AS (sign), and AQ (quotient).

The X input port can be fed by either the AX register set or any result register via the R-bus (AR, MR0, MR1, MR2, SR0, or SR1). The AX register set contains two registers, AX0 and AX1. The AX register can be loaded from the DMD bus. The Y input port can be fed by either the AY register set or the ALU feedback (AF) register. The AY register set contains two registers, AY0 and AY1. The AY registers can be loaded from either the DMD bus or the PMD bus.

The register outputs are dual-ported so that one register can provide input to the ALU while either one simultaneously drives the DMD bus. The ALU output can be loaded into either the AR register or the AF register.

The AR register has a saturation capability; it can be automatically set to plus or minus the maximum value if an overflow or underflow occurs. The AR register can drive both the R bus and the DMD bus and can be loaded from the DMD bus.

The ALU contains a duplicate bank of registers shown in Figure 7.26 behind the primary registers. The secondary set contains all the registers described above (AX0, AX1, AY0, AY1, AF, AR). Only one set is accessible at a time. The two sets of registers allow fast context switching, such as for interrupt servicing.

Example ALU Instructions

- AR = AX0 + AY0
- AF = MR1 XOR AY1
- AR = AX0 + AF

7.11.3. Multiplier–Accumulator

The MAC implements high-speed multiply, multiply–add, and multiply–subtract operations. A block diagram of the MAC section is shown in Figure 7.27.

The multiplier takes two 16-bit inputs, X and Y, and generates one 32-bit output, P. The 32-bit output is routed to a 40-bit accumulator which can add or subtract the P output from the value in MR. MR is a 40-bit register which is divided into three sections: MR0 (bits 0–15), MR1 (bits 16–31), and MR2 (bits 21–29). The result of the accumulator is either loaded into the MR register or into the 16-bit MAC feedback (MF) register. The multiplier accepts the X and Y inputs in either signed or unsigned formats.

In default operation, the result is shifted 1 bit to the left to remove the redundant sign bit for fractional justification; an optional mode on the ADSP-2101 inhibits this shift for integer operations. The accumulator generates one status bit, MV, which is set when the accumulator result overflows the 32-bit boundary. A saturate instruction is available to change the contents of the MR register to the maximum 32-bit value if MV is set. The accumulator also has

FFT HARDWARE IMPLEMENTATION

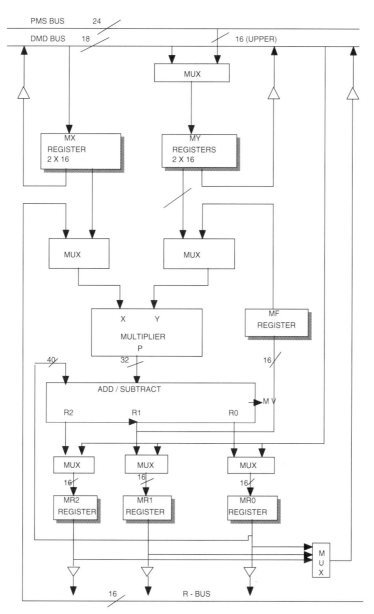

FIGURE 7.27 The MAC block diagram.

the capability for rounding the 40-bit result at the boundary between bit 15 and bit 16.

The MAC and ALU registers are similar. The X input port can be fed by either the MX register set (MX0, MX1) or any result register via the R-bus (AR, MR0, MR1, MR2, SR0, or SR1). The MX register set is readable and loadable from the DMD bus and has dual ported outputs.

The Y input port can be fed by either the MY register set (MY0, MY1) or the MF register. The MY register set is readable from the DMD bus and readable and loadable from both the DMD and PMD bus. Its outputs are also dual ported. The accumulator output can be loaded into either the MR register or the MR register. The MR register is connected to both the R-bus and the DMD-bus. Like the ALU section, the MAC section contains two complete bank registers (MX0, MX1, MY0, MY1, MR, MR0, MR1, MR2) to allow fast context switching.

Example MAC Instructions

- MR = MXO * MY0
- MR = 0
- MRF = AR * MF
- MR = MX0 * MF
- MR = MR + Mx1 * MY0

7.11.4. Shifter

The shifter gives the ADSP-2101 its unique capability to handle data formatting and numeric scaling. Figure 7.28 shows a block diagram of the shifter.

The shifter can be divided into the following components: the shifter array, the OR–PASS logic, the exponent detector, and the exponent compare logic.

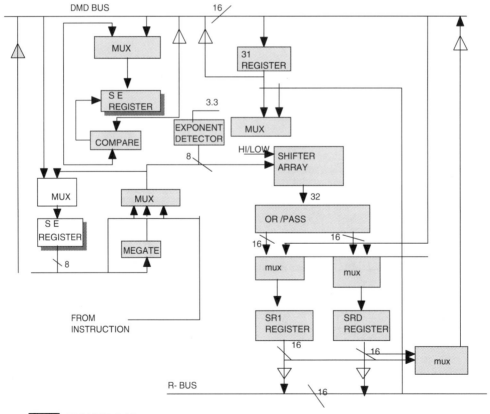

FIGURE 7.28 Shifter block diagram.

These components give the shifter its six basic functions: arithmetic shift, logical shift, normalization, denormalization, derive exponent, and derive block exponent.

The shifter array is a 16 × 32 barrel shifter. It accepts a 16-bit input and can place it anywhere in the 32-bit output field, from off-scale right to off-scale left. The shifter can perform arithmetic shifts (shifter output is sign-extended to the left) or logical shifts (shifter output is zero-filled to the left). The placement of the 16-bit input is determined by the control code C and the HI/LO reference signal.

Example Shifter Instructions

- SR = ASHIFT SI BY −6
- SR = LSHIFT SR BY 3
- SR = NORM MR1

7.11.5. Data Address Generators

A block diagram of a data address generator is shown in Figure 7.29. The data address generators (DAGs) provide indirect addressing for data stored in the program and data memory spaces. The processor contains two independent DAGs so that two data operands (one in program memory and one in data memory) can be addressed simultaneously. The two data address generators are identical except the DAG1 has a bit reversal option on the output (used for FFTs) and can only generate data memory addresses, while DAG2 can generate both program and data memory addresses but has no bit reversal capability. Both DAGs can also be used for serial port autobuffering.

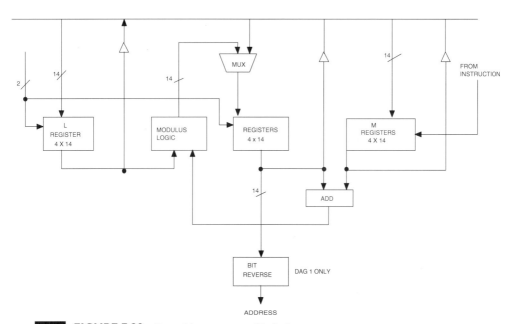

FIGURE 7.29 Data address generator block diagram.

There are three register files in each DAG: the modify (M) register file, the index (I) register file, and the length (L) register file. Each of these register files contains four 14-bit registers that are readable and loadable from the DMD bus. The I register holds the actual addresses used to access external memory. When using the indirect addressing mode, the selected I register content is driven onto either the PMA or DMA bus. This value is post modified by adding the (signed) contents of the selected M register. The modified address is passed through the modulus logic.

Associated with each I register is an L register which contains the length of the buffer addressed by the I register. The L register and the modulus logic together enable circular buffer addressing with automatic wraparound at the buffer boundary. Automatic wraparound is also used by the serial ports to generate the serial port interrupt when operating in autobuffering mode. The modulus logic is disabled by setting the L register to zero.

Example Addressing Instructions

- AX0 = DM (10,M3)
- MODIFY (L1,M2)
- MR = MR + MX0 * MY0 = DM(I0,M1), MY0 = PM(I4,M4)

7.11.6. Program Sequencer

The program sequencer incorporates powerful and flexible mechanisms for program flow control such as zero-overhead looping single-cycle branching (both conditional and unconditional) and automatic interrupt processing. Figure 7.30 shows a block diagram of the program sequencer.

The sequencing logic controls the flow of the program execution. It outputs a program memory address onto the PMA bus from one of four sources: the PC incrementer, PC stack, instruction register, or interrupt controller. The next address source selector controls which of these four sources is selected based on the current instruction word and the processor status. A fifth possible source for the next program memory address is provided by DAG2 when a register indirect jump is executed.

The program counter (PC) is a 14-bit register which contains the address of the currently executing instruction. The PC output goes to the incrementer. The incremented output is selected as the next program memory address if program flow is sequential. The PC value is pushed into the 16×14 PC stack when a CALL instruction is executed or when an interrupt is processed. The PC stack is popped when the return from a subroutine or interrupt is executed. The PC stack is also used in zero-overhead looping.

The program sequencer section contains six status registers. These are the arithmetic status register (ASTAT), the stack status register (SSTAT), the mode status register (MSTAT), the interrupt control register (ICNTL), the interrupt mask register (IMASK), and the interrupt force and clear register (IFC).

The interrupt controller allows the processor to respond to the six possible interrupts with a minimum of overhead. Individual interrupt requests are logically ANDed with the bits in IMASK; the highest priority unmasked interrupt is then selected.

FFT HARDWARE IMPLEMENTATION

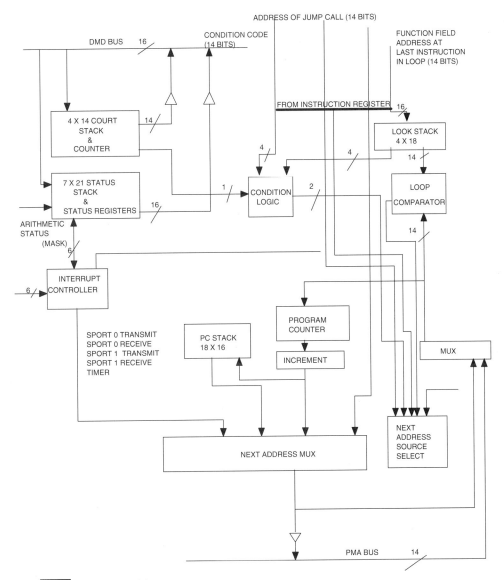

FIGURE 7.30 ADSP-2101 program sequencer.

The interrupt control register, ICNTL, allows each interrupt to be set as either edge or level sensitive. Depending on bit 4 in ICNTL, interrupt routines can either be nested with higher priority interrupts taking precedence or processed sequentially with only one interrupt service active at a time.

7.11.7. Serial Ports

The ADSP-2101 incorporates two complete serial ports (SPORT0 and SPORT1) for serial communications and multiprocessor coordination. A block diagram of one of the serial ports is shown in Figure 7.31. Each serial port has a five-pin interface consisting of the signals shown in Figure 7.31.

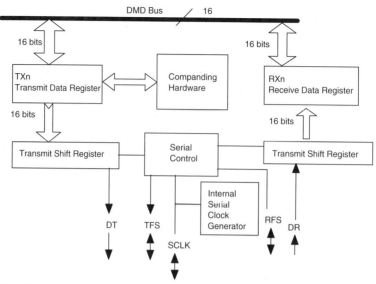

FIGURE 7.31 Serial port block diagram.

Each SPORT has a receive and a transmit register. Companding (a contract of COM pressing and exPANDing) is the process of logarithmically encoding data to reduce the number of bits that must be sent. The ADSP-2101 supports both of the widely used algorithms for companding: A-law and u-law. The type of companding can be independently selected for each SPORT.

7.11.8. System Interface

Figure 7.32 shows a basic system configuration with the ADSP-2101, two serial codecs, a boot EPROM, and optional external program and data memories. Up to 15 K words of data memory and 16 K words of program memory can be supported. Programmable wait-state generation allows the processor to interface easily to slow memories.

The ADSP-2101 also provides one external interrupt and two serial ports or three external interrupts and one serial port.

7.11.9. Interfacing ADCs and DACs to Digital Signal Processors

Interfacing an ADC or a DAC to a fast DSP parallel bus (such as the ADSP-2101) requires an understanding of how the DSP processor reads data from a memory-mapped peripheral (the ADC) and how the DSP processor writes data to a memory-mapped peripheral (to DAC).

A block diagram of a typical parallel DSP interface to an external ADC is shown in Figure 7.33. The diagram has been simplified to show only those signals associated with reading data from an external memory-mapped peripheral device. The timing diagram for the ADSP-2101 read cycle is shown in Figure 7.34.

FFT HARDWARE IMPLEMENTATION

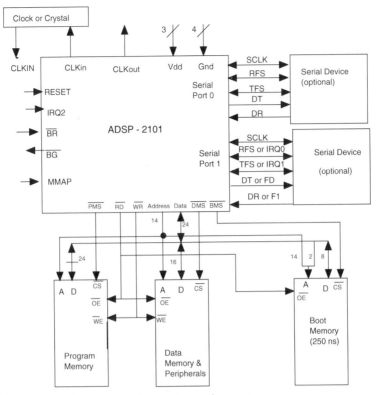

FIGURE 7.32 ADSP-2101 basic system configuration.

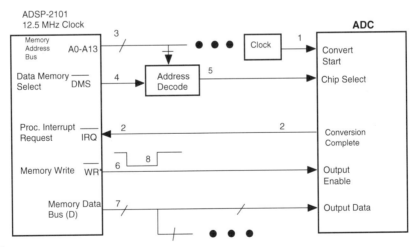

FIGURE 7.33 ADC/ADSP-2101 parallel interface.

The read process starts when the peripheral device (such as an ADC) asserts the processor interrupt request (IRQ) line. The processor then places the address of the peripheral initiating the interrupt request on the memory address bus (A0-A13). At the same time, the processor asserts the data memory select line.

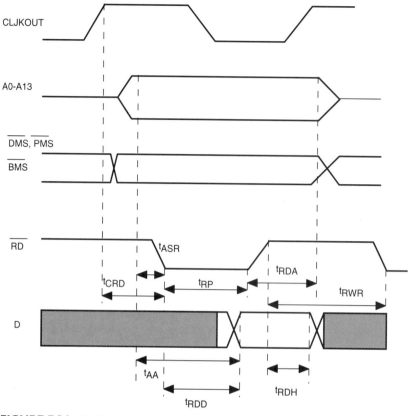

FIGURE 7.34 ADSP-2101 memory read timing.

The ADSP-2101 can easily be interfaced to slow peripheral devices using its programmable wait-state-generation capability. Three registers control wait-state generation for boot, program, and data memory interfaces. You can specify 0 to 7 wait states for each parallel memory interface. Each wait state added increases the allowable external data memory access time by an amount equal to the processor clock period (80 ns for the ADSP-2101 operating at 12.5 MHz). The data memory address, DMS, and RD lines are all held stable for an additional amount of time equal to the duration of the wait states on the peripheral device usually called output enable or read. The rising edge of the $\overline{RD}$ signal is used to clock the data on the data bus into the DSP processor. After the rising edge of the $\overline{RD}$ signal, the data on the data bus must remain valid for t_{RDH} ns, the data hold time. In the case of the ADSP-2101, this value is 0 ns.

The key timing requirements for the peripheral device are shown as follows:

t_{ASR} = address and data memory select setup before READ DATA low

$t_{ASR} = 0.25 t_{CK} - 15$ ns minimum

t_{RDD} = READ DATA low to data valid

t_{CK} = processor clock period

$t_{RDD} = 0.7 t_{CK} - 15$ ns + # wait states $* t_{CK}$ maximum

The t_{RDD} specification determines the peripheral device data access requirement. In the case of the ADSP-2101, $t_{RDD} = 25$ ns minimum. If the access time of the peripheral is greater than this, wait states must be added or the processor speed reduced.

7.11.10. Parallel ADC-to-DSP Interface

The conversion process in a sampling ADC is initiated by a pulse often called the encode command, or start-convert. The leading (or trailing) edge of this pulse tells the internal ADC sample–hold mode that the conversion process can take place. Extreme care must be taken to ensure that this pulse is both jitter-free and noise-free. Any sample-to-sample variation in the occurrence of this edge has the same effect as aperture jitter and will produce a corresponding degradation in the overall ADC signal-to-noise ratio. For this reason, the start-convert signal is usually generated by a stable source external to the DSP processor.

The various timing pulses required to carry out the actual internal conversion process (after receipt of the convert-start command) may be generated in several ways depending upon the individual ADC design. In some ADCs the convert-start pulse triggers an internal oscillator or timing chain which in turn controls the conversion. In other ADCs, a user-supplied asynchronous external clock is required.

At some point in time after the convert-start pulse edge, the internal ADC conversion process is completed. In the case of a parallel-output ADC, a single pulse called data valid, data ready, read data, conversion complete, end-of-conversion, or busy–interrupt is asserted. This pulse is used to drive an interrupt request input of the DSP processor. The DSP then places the address of the ADC on the data memory address bus and asserts the data memory select line, which in turn enables the address decoder. The chip select input to the ADC is then asserted along with the read data line from the DSP. The read data line is connected to the read input of the ADC. Asserting the read line to the ADC enables the tristate parallel outputs which are connected to the data memory data bus. The DSP then reads the ADC data into an internal register on the rising edge of the read data pulse. For the circuit shown in Figure 7.33 to operate properly, the timing between the two devices must be made compatible. This will be illustrated by considering a representative example of the ADSP-2101 processor interfaced to the AD7871 ADC.

The AD 7871 is a 14-bit, 83-kSPS ADC, which can operate in either the parallel or serial mode. A functional block diagram of the AD7871 is shown in Figure 7.35. Specifications for the ADSP-2101 are given for a clock frequency of 12.5 MHz.

Examining the timing specifications shown in Table 7.3 reveals that for the timing between the devices to be compatible, one software wait state must be programmed into the ADSP-2101. A simplified interface diagram for the two devices is shown in Figure 7.36. The conversion complete signal from the AD7871 is designated BUSY/INT.

Parallel interfaces with other DSP processors can be designed in a similar manner by carefully examining the timing specifications for all appropriate signals for each device. The interface between the ADSP-2100 microprocessor

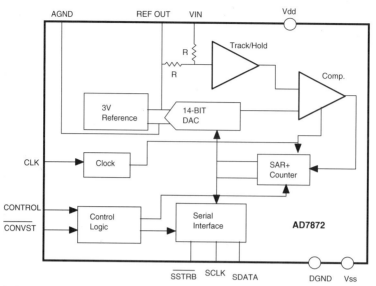

FIGURE 7.35 The AD7871 14-bit ADC functional diagram.

TABLE 7.3 ADSP-2101 and AD7871 parallel read interface timing specifications

ADSP-2101 Processor (12.5 MHz)	AD7871 ADC
t_{ASR} (data address, data memory, select setup time before RD low) = 5 ns min	t_2 (CS to RD setup time) = 0 ns min (must add address decode time to this value)
t_{RP} (RD pulse width) = 30 ns+# wait states * 80 ns min	t_3 (RD pulse width) = 60 ns min
t_{RDD} (RD low to data valid) = 25 ns+# wait states * 80 ns min	t_6 (data access time after RD) = 57 ns max
t_{RDH} (data hold from RD high) = 0 ns min	t_7 (bus relinquish time after RD) = 5 ns min

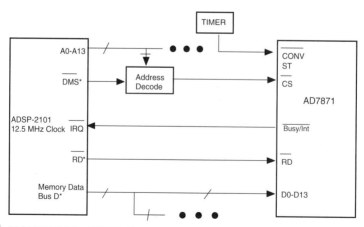

FIGURE 7.36 AD7871 ADC parallel interface to ADSP-2101.

FFT HARDWARE IMPLEMENTATION

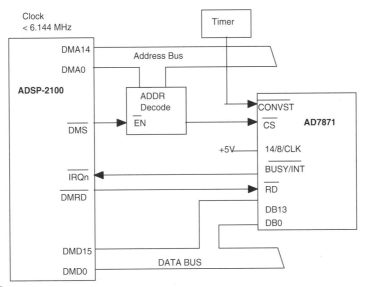

FIGURE 7.37 AD7871 parallel interface to ADSP-2100.

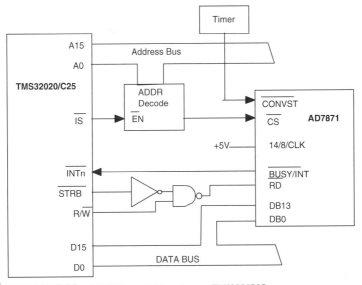

FIGURE 7.38 AD7871 parallel interface to TMS320/C25.

(J-Grade, 6.144-MHz clock) is shown in Figure 7.37. Interfacing the AD7871 ADC to faster versions of the ADSP-2100 series requires the addition of wait states using the data memory acknowledge (DMACK) signal. The DMACK signal indicates that the memory-mapped peripheral is ready for data transfer. If DMACK is not asserted when checked by the processor, wait states are automatically generated until DMACK is asserted.

The parallel interface between the AD7871 and the TMS32020/C25 is shown in Figure 7.38.

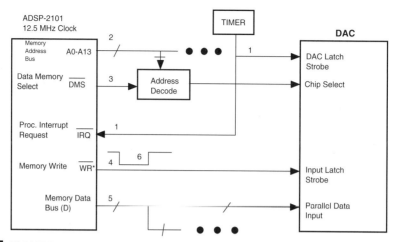

FIGURE 7.39 DAC–ADSP-2101 parallel interface.

7.11.11. Parallel Interfacing to DSP Processors: Writing Data to Memory-Mapped DACs

A simplified block diagram of a typical DSP interface to a peripheral device showing write-mode signals is shown in Figure 7.39. The memory-write cycle timing diagram for the ADSP-2101 is shown in Figure 7.40. The write process may be initiated by the peripheral device by asserting the DSP interrupt request line indicating that the peripheral is ready to accept a new parallel data word. The DSP then places the address of the peripheral device on the address bus and asserts the data memory select ($\overline{DMS}$) line. This causes the output of the address decoder to assert the chip select input to the peripheral. The write ($\overline{WR}$) output of the DSP is asserted t_{ASW} ns after the negative-going edge of the signal. The width of the $\overline{WR}$ pulse is t_{WP} ns. Data are placed on the data bus (D) and are valid t_{DW} ns before the $\overline{WR}$ line goes high. The positive-going transition of the $\overline{WR}$ line is used to clock the data on the data bus (D) into the external parallel memory. The data on the data bus remain valid t_{DH} ns after the positive-going edge of the signal.

The ADSP-2101 can easily be interfaced to slow peripheral devices using its programmable wait-state-generation capability, which causes the memory address, DMS, WR, and data output lines to remain stable for an amount of additional time equal to the duration of the wait states.

7.11.12. Parallel DAC-to-DSP Interface

A typical parallel interface between a DSP and a DAC is shown in Figure 7.41. In most DSP applications, the DAC is operated continuously from a stable clock source which is external to the DSP processor. The DAC should have double buffering: an input latch to handle the asynchronous DSP interface, and a second latch that drives the DAC current switches. The DAC latch strobe is derived from the external stable clock. In addition to clocking the DAC latch, the DAC strobe is also used to generate a processor interrupt that indicates the

FFT HARDWARE IMPLEMENTATION

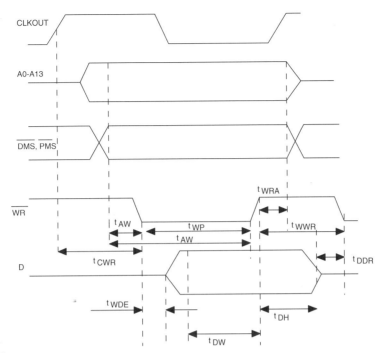

FIGURE 7.40 ADSP-2101 memory write timing.

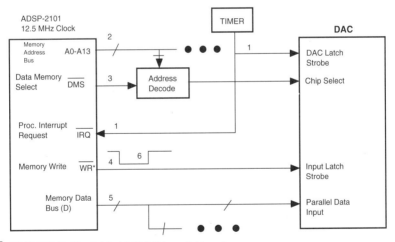

FIGURE 7.41 DAC–ADSP-2101 parallel interface.

DAC is ready for new input data. The processor then asserts the data memory select line and places the DAC address on the memory address bus. The DAC chip select is then asserted, and the data memory write line loads the next data word on the data memory data bus into the DAC input latch. This completes the write cycle, and the DAC is now ready to receive the next DAC latch strobe from the external source. For the circuit shown in Figure 7.41 to operate properly, the timing between the two devices must be made compatible. This

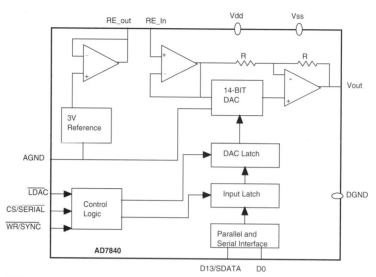

FIGURE 7.42 AD7840 14-bit DAC functional diagram.

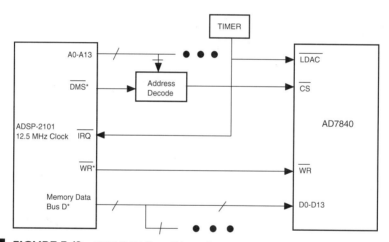

FIGURE 7.43 AD7840 DAC parallel interface to ADSP-2101.

will be illustrated by considering a representative example of the ADSP-2101 processor interfaced to the AD7840 DAC.

The AD7840 is a 14-bit 100-kSPS DAC that has both parallel and serial interface capability. A block diagram of the device is shown in Figure 7.42. A simplified interface diagram for the two devices is shown in Figure 7.43.

Parallel interfaces with other DSP processors can be designed in a similar manner by carefully examining the timing specification for all appropriate signals for each device. The interface between the ADSP-2100 microprocessor (clock speeds up to 8.192 MHz) is shown in Figure 7.44. Interfacing the AD7840 to the ADSP-2100 at clock speeds of greater than 8.144 MHz requires the addition of wait states using the ADSP-2100 DMACK signal. The parallel

FFT HARDWARE IMPLEMENTATION

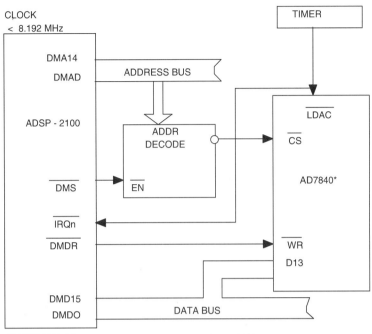

FIGURE 7.44 AD7840 parallel interface to ADSP-2100.

interface between the AD7840 DAC and the TMS32020–C25 is shown in Figure 7.45.

7.11.13. Serial Interfacing to DSP Processors

DSP processors that have serials ports (such as the ADSP-2101, DSP56000, and the TMS32020–C25) provide a simple interface to peripheral ADCs and DACs. Use of the serial port eliminates the need for using large parallel buses to connect the ADCs and DACs to the DSP. To understand serial data transfer better, we will first examine the serial port operation of the ADSP-2101.

A block diagram of one of the two serial ports of the ADSP-2101 is shown in Figure 7.46. The transmit (TX) and receive (RX) registers are identified by name in the ADSP-2101 assembly language, not memory mapped.

In the receiving portion of the serial port, the receive frame synchronization (RFS) signal initiates reception. The serial receive data (DR) from the external device (ADC) is transferred into the receive shift register 1 bit at a time. The negative-going edge of the serial clock (SCLK) is used to clock the serial data from the external device into the receive shift register. When a complete word has been received, it is written to the receive register (RX), and the receive interrupt for that serial port is generated. The receive register is then read by the processor.

Writing to the transmit register readies the serial port for transmission. The transmit frame synchronization (TFS) signal initiates transmission. The value in

7 HARDWARE APPROACH TO DIGITAL SIGNAL PROCESSING

AD7840 PARALLEL INTERFACE TO TMS3202

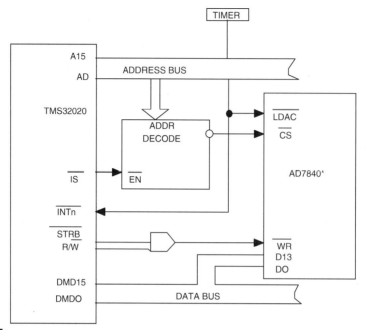

FIGURE 7.45 AD7840 parallel interface to TMS32020.

ADSP-2101 SERIAL PORT BLOCK DIAGRAM

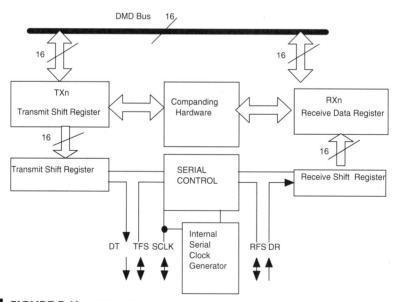

FIGURE 7.46 ADSP-2101 serial port block diagram.

FFT HARDWARE IMPLEMENTATION

the transmit register (TX) is then written to the internal transmit shift register. The data in the transmit shift register are sent to the peripheral device (ADC) 1 bit at a time, and the positive-going edge of the serial clock (SCLK) is used to clock the serial transmit data (DT) into the external device. When the first bit has been transferred the serial port generates the transmit interrupt. The transmit register can then be written with new data, even though the transmission of the previous data is not complete.

In the normal framing mode, the frame sync signal (RFS of TFS) is checked at the falling edge of SCLK. If the framing signal is asserted, data are available (transmit mode) or latched (receive mode) on the next falling edge of SCLK. The framing signal is not checked again until the word has been transmitted or received. In the alternate framing mode, the framing signal is asserted in the same SCLK cycle as the first bit of a word. The data bits are latched on the falling edge of SCLK, but the framing signal is checked only on the first bit. Internally generated framing signals remain asserted for the length of the serial word. The alternate framing mode of the serial port in the ADSP-2101 is normally used to receive data from ADCs and transmit data to DACs.

7.11.14. Serial ADC-to-DSP Interface

A timing diagram of the ADSP-2101 serial port operating in the receive mode (alternate framing) is shown in Figure 7.47. The first negative-going edge of the SCLK to occur after the rising edge of the RFS input clocks the MSB data from the ADC into the serial input latch. The key timing specifications of concern are the serial data setup (t_{SCS}) and hold times (t_{SCH}) with respect to the negative-going edge of the SCLK. In the case of the ADSP-2101, these values are both 10 ns minimum. The RFS setup and hold times are also 10 ns, respectively. Most peripheral ADCs will have no trouble meeting these specifications, even at the maximum serial data transfer rate of 12.5 MHz.

The AD7872 ADC is a 14 bit, 83-kSPS serial-only version of the AD7871. A block diagram of the device is shown in Figure 7.48. The device operates on a 2-MHz external or internal clock. Figure 7.49 shows the AD7872 interfaced to the ADSP-2101. The timing diagram of the AD7872 is shown in Figure 7.50.

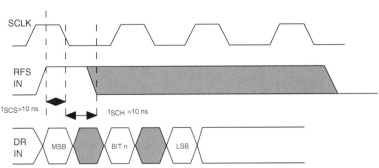

FIGURE 7.47 ADSP-2101 serial port receiving timing.

226   7 HARDWARE APPROACH TO DIGITAL SIGNAL PROCESSING

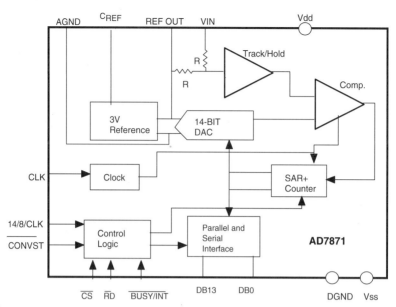

FIGURE 7.48 AD7872 14-bit output ADC block diagram.

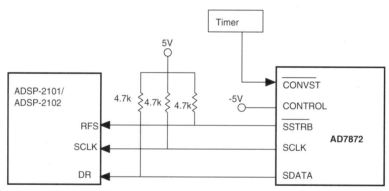

FIGURE 7.49 AD7872 serial interface to ADSP-2101.

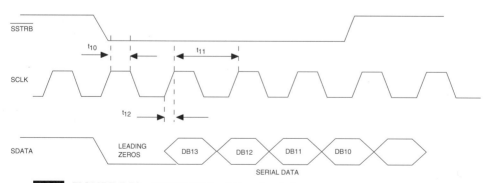

FIGURE 7.50 AD7872 ADC serial interface timing.

FFT HARDWARE IMPLEMENTATION

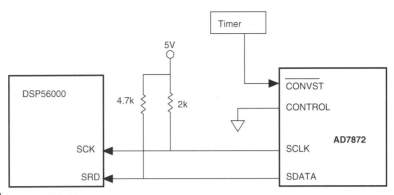

FIGURE 7.51 AD7872 serial interface to DSP56000.

The SSTRBbar signal is active-low, so the ADSP-2101 must be programmed to accept an inverted RFS input. The serial clock operates at a frequency of 2 MHz (500-ns period). The serial clock can be programmed for either continuous or gated operation. In this example, it operates in the continuous mode. The data bits are valid t_{12} ns (155-ns max) after the positive-going edges of SCLK. This allows a setup time of $250 - 155 = 95$ ns minimum before the negative-going edges of SCLK. The hold-time after the negative-going edge of SCLK is therefore at least equal to one-half the clock period, or 250 ns. The positive-going edge of the SSTRB signal occurs t_{13} ns (140 ns max) after the positive-going edge of SCLK after the last data bit is transferred. This allows $250 - 140 = 110$ ns minimum before the next negative-going edge of SCLK. These simple calculations show that data and RFS setup and hold requirements of the ADSP-2101 (10 ns) are met with considerable margin.

The ADSP-2101 can be easily programmed to generate the 2-MHz serial clock for the AD7872 if desired. Details can be found in the ADSP-2101 *User's Manual–Architecture*. The convert start (CONVST) signal is generated externally to the AD7872 from a stable clock source which is asynchronous to the serial clock.

The serial interface between the AD7872 and the DSP56000 is shown in Figure 7.51, and the interface with the TMS32020–C25 is shown in Figure 7.52. The simple interfaces shown to the three DSP processors are referred to as zero-chip interfaces because no additional glue logic is required.

7.11.15. Serial DAC-to-DSP Interface

A timing diagram of the ADSP-2101 serial port operating in the alternate framing transmit mode (with internally generated transmit frame sync) is shown in Figure 7.53. The first negative-going edge of the SCLK to occur after the rising edge of the TFS output clocks the MSB data from the serial port into the DAC serial input latch. The process continues until all serial bits have been transferred into the DAC serial input latch. The key timing specifications of concern are the data output setup and hold times with respect to the negative-going edge of the SCLK. The ADSP-2101 specifies that the TFS output will be a valid high t_{RH} ns (15 ns max) after the positive-going edge of the SCLK. The serial

AD7872 SERIAL INTERFACE TO TMS32020/C25

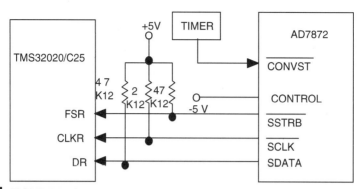

FIGURE 7.52 AD7872 serial interface to TMS32020–C25.

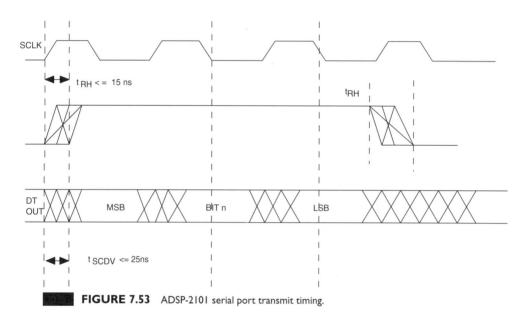

FIGURE 7.53 ADSP-2101 serial port transmit timing.

transmit data are valid t_{SCDV} ns (25 ns max) after the positive-going edge of SCLK. Due to the high speed of the serial port interface of the ADSP-2101, data setup and hold times are approximately equal to one-half the period of the serial clock for clock rates up to 12.5 MHz.

The AD766 is a 16-bit serial DAC that can operate at sample rates up to 500 SPS and is fully specified in terms of both dc and ac parameters such as THD and SNR. A block diagram of the device is shown in Figure 7.54. Data are transmitted to the AD766 in a bit stream composed of 16-bit words with a serial, MSB first format. Three signals must be present to achieve proper operation: the data, clock, and latch enable signals. Input data bits are clocked into the input register on the falling edge of the clock signal. The LSB is clocked in on the 16th clock pulse.

FFT HARDWARE IMPLEMENTATION

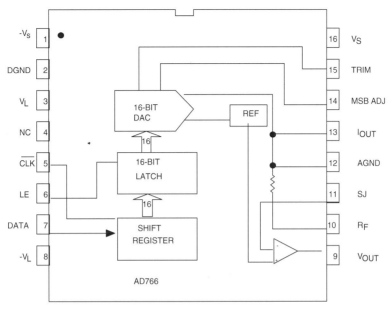

FIGURE 7.54 AD766 16-bit DAC functional diagram.

When all data bits are loaded, a low-going latch enable pulse updates the DAC input. Figure 7.55 illustrates the general signal requirements for data transfer for the AD766. Data setup and hold times (with respect to the negative-going SCLK edge) are each 15 ns.

The negative-going edge of the latch enable must occur at least 15 ns before the negative-going edge of SCLK. These detailed timing requirements are illustrated by Figure 7.56. These time requirements are compatible with the serial ports of popular DSP processors. The AD766 input clock can run at a 12.5-MHz rate. This clock rate will allow sampling rates up to 500 kSPS.

The SDSP-2101 incorporates two complete serial ports that can be directly interfaced to the AD766, as shown in Figure 7.57. Using both serial ports, two AD766s can be directly interfaced with no additional hardware. The zero-chip interface to the TMS32020–C25 is shown in Figure 7.58. The maximum serial clock rate for the TMS32020–C25 is 5 MHz. Figure 7.59 shows the serial interface to the DSP56000–56001.

7.11.16. Interfacing I–O Ports and CODECs to DSPs

Since most DSP applications require both an ADC and a DAC, I–O ports and CODECs have been developed that integrate the two functions on a single chip as well as providing easy-to-use interfaces to standard DSPs.

The AD7869 is a 14-bit I–O port, which is functionally equivalent to the AD7868. These devices are fully specified in terms of ac and dc performance.

7 HARDWARE APPROACH TO DIGITAL SIGNAL PROCESSING

AD766 DAC SIGNAL REQUIREMENTS

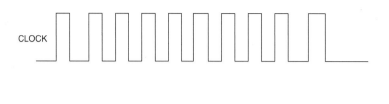

FIGURE 7.55 AD766 DAC signal requirements.

AD766 TIMING REQUIREMENTS

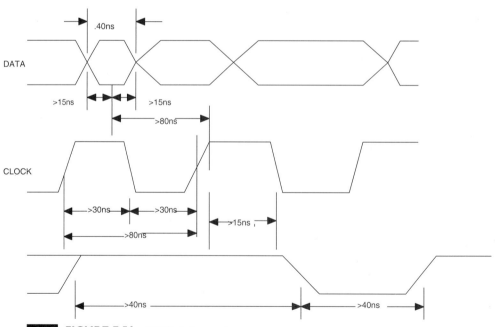

FIGURE 7.56 AD766 timing requirements.

FFT HARDWARE IMPLEMENTATION

AD766 SERIAL INTERFACE TO ADSP-2101

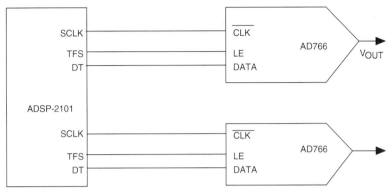

FIGURE 7.57 AD766 serial interface to ADSP-2101.

AD766 SERIAL INTERFACE TO TMS32020–C25

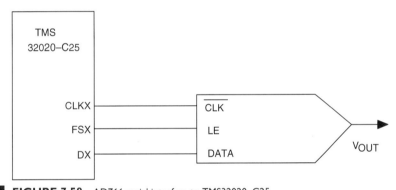

FIGURE 7.58 AD766 serial interface to TMS32020–C25.

AD766 SERIAL INTERFACE TO DSP5600

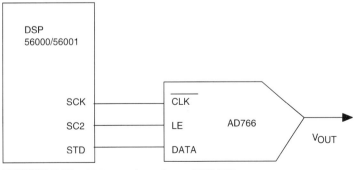

FIGURE 7.59 AD766 serial interface to DSP56000.

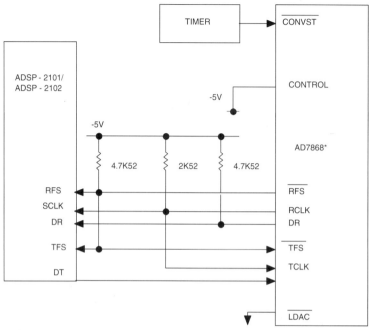

FIGURE 7.60 AD7868 I–O port interface to ADSP-2101.

The SNR (including distortion) of the AD7868 is 72 dB, while the AD7869 is 82 dB. Both devices provide simple interfaces to the serial ports of standard DSP microcomputers such as the ADSP-2101 (see Figure 7.60), TMS32020–C25, and the DSP56000.

The ADSP-28msp02 is a complete voiceband CODEC (ADC and DAC) based on sigma-delta technology. A block diagram is shown in Figure 7.61. The device provides a complete analog front-end for high-performance voiceband DSP applications.

Compared to traditional m-law which is well known, the ADSP-28msp02's linear coded ADC and dc maintain wide dynamic range throughout the transfer function. The encoder side of the device consists of two selectable analog input amplifiers and a sigma-delta ADC. The gain of input amplifiers can be adjusted with the use of external resistors from -12 to $+26$ dB. An optional 20-dB preamplifier can be inserted before the ADC. The preamplifier and the multiplexer are configured by bits in the control register. The decoder consists of a sigma-delta DAC and a differential amplifier. The output of the DAC drives an analog smoothing filter which converts the data into an analog voltage. The gain of the smoothing filter and PGA can be adjusted via the control register from -15 to $+6$ dB in 3-dB steps. The ADSP-28msp02 easily interfaces to the serial ports of popular DSP microcomputers such as the ADSP-2101 as shown in Figure 7.62.

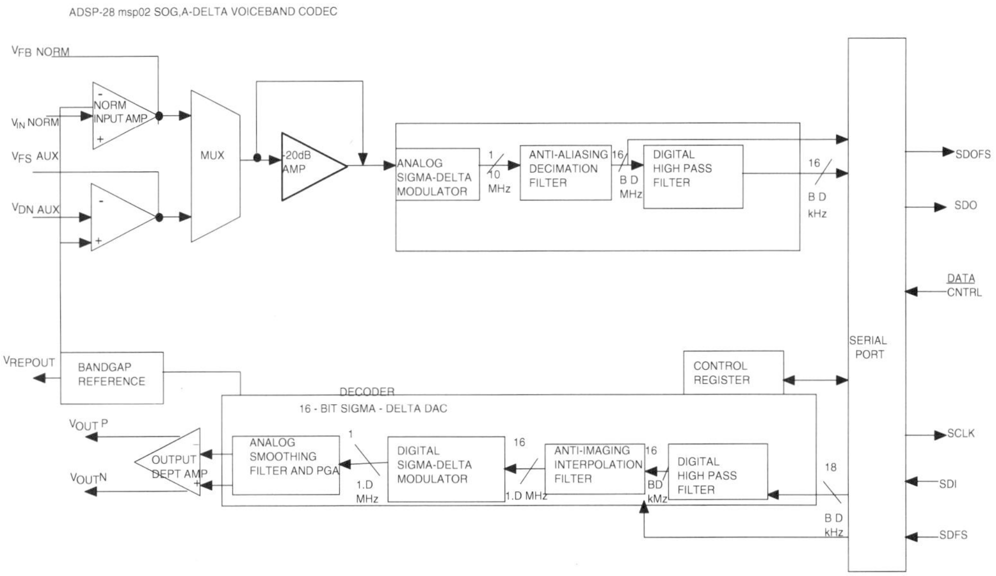

FFT HARDWARE IMPLEMENTATION

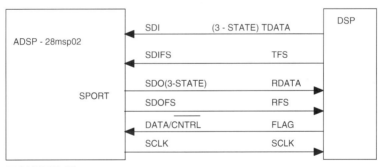

FIGURE 7.61 ADSP-28msp02 sigma-delta voiceband CODEC.

FIGURE 7.62 ADSP-28msp02 CODEC serial port DSP interface.

7.11.17. Serial versus Parallel DSP Interface Summary

Some DSP processors such as the ADSP-2100 support only memory-mapped peripherals and have no serial port. A large number of peripheral devices can be connected to the parallel address and data bus. Each device is treated as a single location in the data memory. A number of high-performance ADCs and DACs are available with parallel interfaces. Data setup and hold specifications write and read pulse widths, etc. must be examined carefully to ensure that there are no interface timing violations. Conflicts frequently occur because DSP processors are designed to operate at clock frequencies often exceeding 10 MHz, while ADCs and DACs used in most DSP applications rarely exceed sampling rates of 500 kSPS. These interfacing timing conflicts can usually be resolved with the addition of software or hardware wait states.

ADCs that interface to parallel DSPs must have tristate outputs so that the data bus can be shared among other peripherals. The convert-start signal for the ADC is generated externally to minimize jitter. The conversion-complete signal is typically used to generate an interrupt request to the DSP processor. Care must be taken in the routing of the ADC parallel digital outputs to prevent digital switching noise from coupling into the ADC analog input.

7.12. PRACTICAL USE OF DSP: DSP HELPS THE HEARING IMPAIRED

In a normally functioning ear, the ear receives the acoustic pressure waves. The outer ear directs these waves through the ear canal up to the middle ear. In the middle ear, the eardrum converts these sound waves into mechanical vibrations that are conducted by very small bones in the middle ear. The motion of these bones causes movement of the fluid contained in the cochlea, the snail-shaped inner ear. Pressure variations within the fluids in the cochlea displace the flexible basilar membrane. This displacement carries the information about the frequency and amplitude of the acoustic signal. Thousands of tiny hairlike receptors attached to the basilar membrane bend according to the membrane displacement. The bending of the hairs causes the release of an electrochemical substance that causes the neurons to fire. The neurons (through the auditory nerve) transmit this information to the central nervous system where it is interpreted as sound.

In many people suffering from hearing loss, the hair cells in the cochlea are damaged; therefore, there is no way for acoustic pressure waves to be transformed into nerve impulses. A cochlear implant works by bypassing the damaged hair cells and directly stimulating the neurons. Cochlear implants are mainly for those people who cannot benefit from hearing aids. A cochlear implant is an electronic device that electrically stimulates the auditory nerve fibers in the inner ear. A cochlear implant consists of a speech processor, a microphone, a cochlear stimulator, and an electrode array (Figure 7.63).

The microphone at the ear opening (on the headpiece) is similar to those in hearing aids. When the sound is received, it is converted to an electrical signal, and it is sent to the speech processor via a thin cable. The person wears the speech processor as if it were a pager device. The speech processor converts the electrical signal into a particular code based on certain DSP algorithms. The processor sends this information to the headpiece. The headpiece uses radio waves to send these signals to a tiny coil within the cochlear implant or receiver. The receiver stimulator decodes the signal and sends it to the electrode array within the cochlear implant. The electrodes stimulate the nerve fibers.

The cochlea normally responds to multiple frequencies at different locations along the basilar membrane. Cochlear implants can be developed such that multichannel implants stimulate nerve fibers at multiple locations along the cochlea. The DSP algorithm plays a central role in the stimulation method since the neural excitation spreads symmetrically from the source, constraining the frequency coding resolution. The loudness of the sound is dependent on the number of nerve fibers activated and their rates of firing. This means

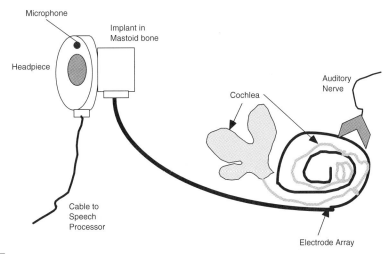

FIGURE 7.63 An illustration of a cochlear implant.

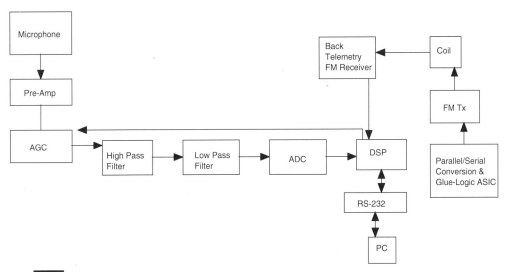

FIGURE 7.64 Block diagram of the cochlear implant.

that the DSP processor can control the sound's loudness by varying the stimulus current's amplitude.

The main objective of the cochlear implant DSP is to split the speech input signals into frequency bands to deliver it to the appropriate electrodes in the cochlea. To sort out all the frequencies an FIR or IIR filter can be used. The FIR filter is more stable but uses more power; the IIR filters are less stable but use less power. A block diagram of the cochlear implant is shown in Figure 7.64.

The preamplifier amplifies the analog signal generated by the microphone. The signal goes through automatic gain control to avoid overdriving the circuitry and clipping the input signal. A high-pass antialiasing filter and a low-frequency preemphasis stage will limit the signal bandwidth to the human-speech frequencies. The ADC hands off the signal to the DSP processor.

In addition to handling the speech-coding strategies and frequency separations, the DSP samples each speech level and accordingly adjusts the automatic gain control (AGC) levels. The speech processor converts the DSP output to a serial bit stream, frequency modulates it, and feeds it to the coil. The coil couples in an inductive manner the RF signal to the implanted receiver and electrode array. The coil in the receiver picks up the frequency-modulated carrier and rectifies it to provide dc power for the implant circuits. The output is capacitively coupled to the electrodes that stimulate the cochlear nerves to prevent dc current flow.

REFERENCES

ADSP-2100 User's Manual and Architecture. Analog Devices.
ADSP-2101/2102 User's Manual/Architecture. Analog Devices.
Eidi, Fares. "An optimized radix-4 fast Fourier transform (FFT)," Analog Devices Application Note E1329-5-9/89.
High Speed Design Seminar. 1990. Analog Devices.
Karagozyan, Kapriel. "Wait state generation on the ADSP-2100 and the ADSP-2100A." Analog Devices Application Note E1317-8-8/89.
Mar, Amy, Editor. 1990. *Digital Signal Processing Applications Using the ADSP-2100 Family.* Prentice-Hall, Englewood Cliffs, NJ.
Rabiner, L. R., and B. Gold, 1975. *Theory and Application of Digital Signal Processing.* Prentice Hall.
Williams, C. S. 1986. *Designing Digital Filters.* Prentice-Hall, Englewood Cliffs, NJ.

8
OPTICAL SENSORS

There are basically two technologies that are widely used in optical sensors, charge-coupled devices (CCDs) and fiber optics. Both are major contributors to the development of advanced diagnostics and therapeutic techniques. In this chapter, we first address CCD and then fiber optics.

8.1. CHARGE-COUPLED DEVICES

The CCD refers to a semiconductor architecture in which a charge is transferred through storage areas. The CCD architecture has three basis steps: (1) charge collection, (2) charge transfer, and (3) the conversion of charge into a measured voltage. The basic building block of the CCD is the metal semiconductor capacitor (MIS), also known as the gate. The most common MIS is the metal oxide semiconductor.

Charge generation is often considered as the initial goal of CCD. Silicon can create an electron–hole pair for each absorbed photon. The electrons and holes can be stored or transformed; charge generation occurs under an MOS capacitor. The charge created at a pixel site is proportional to the incident light. The aggregate effect of all the pixels is to produce a spatially sampled representation of the continuous picture. Pixel readout occurs by sensing the charge transfer between the capacitor at the pixel site.

When an absorbed photon creates an electron–hole pair, photodetection has occurred. The generated carriers must be stored at a site. The absorption coefficient is wavelength specific and decreases with increasing wavelength.

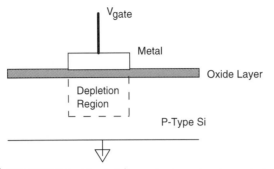

FIGURE 8.1 The depletion region in semiconductor devices.

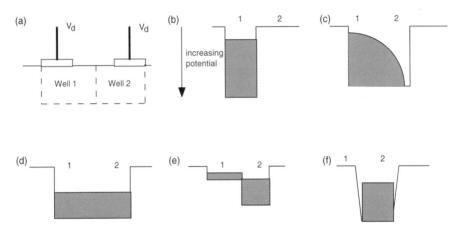

FIGURE 8.2 Movement of electrons in a potential well due to photoelectrons.

Applying a positive voltage to the CCD causes the mobile positive holes in *p*-type silicon to migrate toward the ground electrode because like charges repel. This region, which is devoid of positive charges, is the depletion region. A photon with an energy greater than the energy gap is absorbed in the depletion region producing an electron–hole pair as shown in Figure 8.1.

The electron stays within the depletion region, whereas the hole moves to the ground electrode. The number of electrons collected is proportional to the applied voltage, oxide thickness, and gate electrode area.

The CCD register consists of a series of gates. The manipulation of the gate voltage in a systematic and sequential manner transfers the electrons from one gate to the next in a conveyor-belt-like fashion. For charge transfer, the depletion region must overlap. The depletion regions are gradients, and the gradients must overlap for charge transfer to occur.

Initially, a voltage is applied to gate 1 and photodetectors are collected in well 1 (Figure 8.2). When a voltage is applied to gate 2, electrons move to well 2 in a waterfall manner (c).

The process is rapid and then the charge quickly equilibrates in two wells (d). As the voltage is reduced on gate 1, the well potential decreases

CHARGE-COUPLED DEVICES

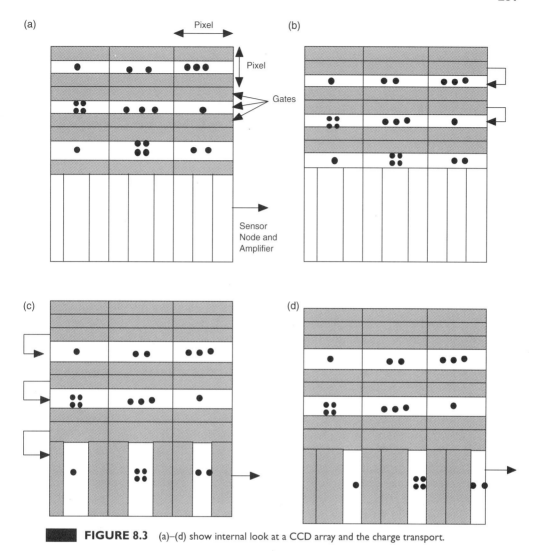

FIGURE 8.3 (a)–(d) show internal look at a CCD array and the charge transport.

and electrons again flow in a waterfall manner into well 2 (e). The process is repeated many times until the charge is transferred through the shift register.

The CCD array is a series of column registers (Figure 8.3). The charge is kept within rows or columns by channel stops and the depletion regions overlap in one direction only. At the end of each column is a horizontal register of pixels. The register accumulates a line at a time, and later it transports the charge packets in a serial mode to an output amplifier. The horizontal serial register must be clocked out to be a sense node before the next line enters the serial register. Therefore, separate vertical and horizontal clocks are required for all CCD arrays. The process creates a serial data stream that represents the two-dimensional image.

Although any number of transfer gates per detector can be used, the number generally varies from two to four. With a three-phase system the charge is stored under one or two gates. This is shown in Figure 8.4. Only 33% of the pixel area is available for well capacity. With equal potential wells, a minimum number of three phases are required to clock out charge packets efficiently.

As the voltage is applied, the charge packet moves to that well. By sequentially varying the gate voltage, the charge moves off the horizontal shift register and onto a sense capacitor. The clock signals are identical (only one master clock is required to drive the array) for all three phases but offset in time (phase).

8.1.1. CCD Arrays

Array architecture is dependent on the application. Full-frame arrays are used in scientific applications. Interline transfer devices are used in consumer product. In Figure 8.5 we can illustrate a full-frame (FFT) array. After integration the image pixels are read out line-by-line through a serial register which clocks its content onto the output sense node. All the charge must be clocked out of the serial register before the next time line can be transferred.

During the reading process, the pixels are continually illuminated, which can result in a smeared image in the direction charge flow. Data rates are limited by the amplifier bandwidth and also by the capabilities of the analog-to-digital converter. A large array can be divided into subarrays that are read out simultaneously, the effective clock rate increases by the number of subarrays. Figure 8.6 illustrates a large array subdivided into four subarrays. Software can reconstruct the original image, where the serial data are devoted and reformatted by a video processor.

A frame transfer image contains two almost identical arrays. One array is used for image pixel and the other one for storage. The storage cells are identical to the light-sensitive cells but are covered with a metal light shield to prevent any light exposure. Once the integration cycle is complete the charge is transferred quickly from the light-sensitive pixels to the storage cells. The transfer time is around 500 μs. The smear is limited only to the time it takes to transfer the image to the storage area.

8.1.2. Interline Transfer

The interline transfer array consists of photodiodes separated by vertical registers that are covered by an opaque metal shield (see Fig. 8.7). After integration, the charge generated by the photodiodes is transferred to the vertical CCD registers very fast and the smear is then minimized. The main advantage of the interline transfer is that the transfer from the active sensors to the shield storage is quick. The shields act like a venetian blind that obscures half the information that is available in the scene. The area fill factor can be as low as 20%. Since the detector is only 20% of the pixel area, the output voltage is only 20% of a detector, which would fill the pixel area.

A fraction of the light can leak into vertical registers but this can be minimized by using shielded storage array. There are several types of transfer register architectures. In Figure 8.8 we can observe a four-phase transfer register that

CHARGE-COUPLED DEVICES

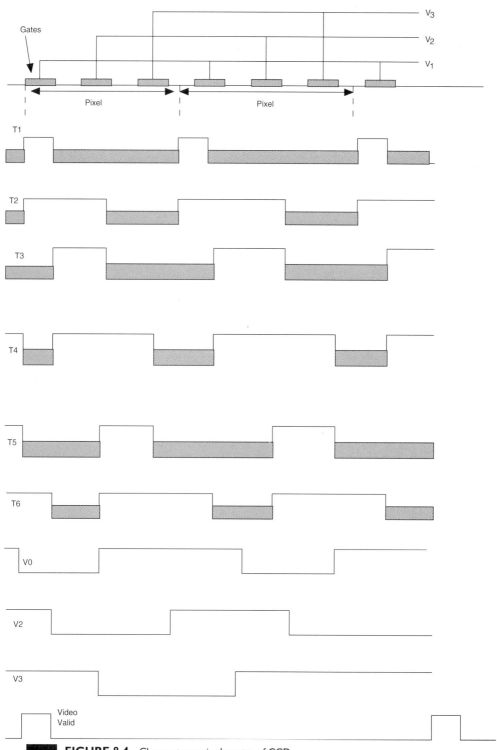

FIGURE 8.4 Charge storage in the gates of CCD.

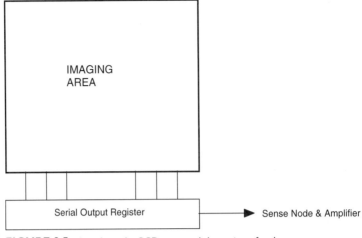

FIGURE 8.5 Interface of a CCD array and the register for data output.

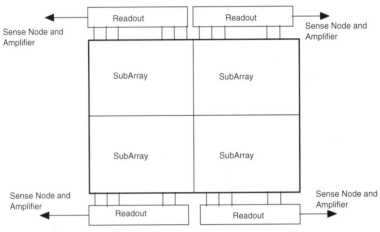

FIGURE 8.6 CCD subarrays and hardware registers for reading data out.

carries charge under two gates. The 2:1 interlace has both fields collected at the same time but are read out alternately.

8.2. OPTICAL FIBER

Communications using light as a signal carrier and optical fibers as the media are termed *optical fiber communications*. The applications of optical fiber communications have increased very rapidly since the first commercial installation of a fiber-optic system in 1977. Today every major telecommunications company spends millions of dollars developing and utilizing an optical fiber communications system. Fiber-optic communication systems can process information using either digital or analog modulation schemes. In the most commonly used fiber-optic communication system, voice, video, and data are converted to a coded

OPTICAL FIBER 243

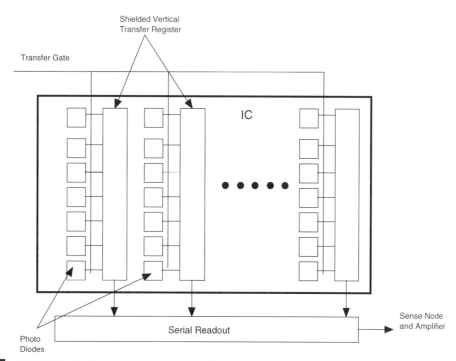

FIGURE 8.7 The interline transfer in a CCD.

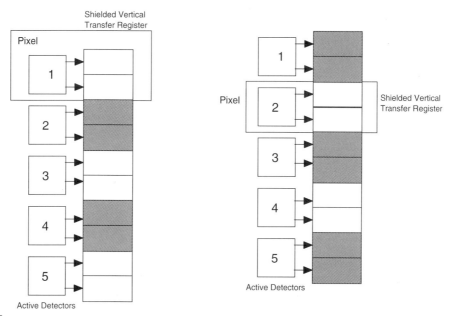

FIGURE 8.8 Detailed layout of the 2:1 interlaced array: (a) the odd field is clocked into the vertical transfer register and (b) the even field is transferred.

pulse stream of light using a suitable light source. This pulse stream is carried by optical fibers to a regenerating or receiving station. At the final receiving station the light pulses are converted into the form of the original information. Optical fiber communication systems are currently used in the fields of telecommunications, data communications, and cable television. These systems are also used for many military, automotive, medical, and industrial applications.

Since the dawn of history, people have used light as a vehicle to carry information. Lanterns on ships and smoke signals or flashing mirrors on land are early examples of how light was used to communicate. It was over a hundred years ago that Alexander Graham Bell (1880) transmitted a telephone signal a distance greater than 200 m using light as a signal carrier. Bell called his invention a "photophone" and obtained a patent for it (Bruce 1973). Bell, however, wisely gave up the photophone in favor of the electric telephone. The photophone, at the time of its invention, could not be exploited commercially because of two basic drawbacks: (1) the lack of a reliable light source and (2) the lack of dependable transmission medium.

Modern light-wave communication systems had their birth in the 1960s. The first demonstration of the ruby laser in 1960 (Maiman 1960) and a demonstration of laser operations in semiconductor devices in 1962 (Hall *et al*. 1962; Nathan *et al*. 1962) were early stepping stones that led to the continuous operation of room temperature, long-lifetime, semiconductor lasers that are in common use today. The laser made available a coherent optical frequency carrier of enormous communication capacity. If a communication system was built that utilized only 0.01% of the laser carrier frequency, its modulation bandwidth would be 30 GHz. In 1966 a parallel evolution of fiber technology was taking place. Although the best existing fibers at that time had attenuation greater than 1000 dB/km, researchers at Standard Telecommunication Laboratories (STC) in England (Kao and Hockham 1966) speculated that losses as low as 20 dB/km should be achievable and they further suggested that such fibers would be useful in telecommunication applications. And they were correct. In 1970 workers at Corning Glass Works (Kapron *et al*. 1970) produced the first fiber with loss under 20 dB/km. Since that time, fiber technology has advanced to the point of producing fibers with loss less than 0.25 dB/km at 1.55 μm. These fibers are approaching the Rayleigh scattering limit of the glass being used to fabricate them. The biggest advantage of an optical fiber communication system is its tremendous information-carrying capacity. There are already many systems that can carry several thousand simultaneous conversations over a pair of optical fibers that are thinner than a human hair.

8.2.1. Classification and Features of Optical Fibers

Fibers that are used for optical communications are waveguides made of transparent dielectrics whose function is to guide light over long distances. An optical fiber consists of an inner cylinder of glass called the core, surrounded by a cylindrical shell of glass of lower refractive index, called the cladding. Optical fibers may be classified in terms of the refractive index profile of the core and whether one mode (single-mode fiber) or many modes (multimode fiber) are propagating

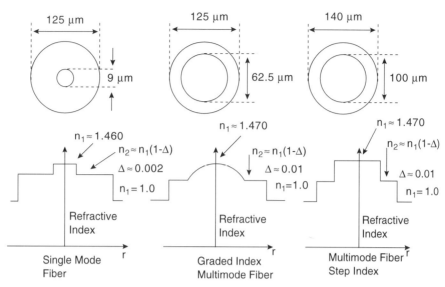

FIGURE 8.9 Dimensions and refractive indexes for commonly used optical fibers.

in the guide (Figure 8.9). If the core, which is typically made of high-silica-content glass or a multicomponent glass, has a uniform refractive index n_1, it is called a *step-index* fiber. If the core has a nonuniform refractive index that gradually decreases from the center toward the core–cladding interface, the fiber is called a *graded-index* fiber. The cladding surrounding the core has a uniform refractive index n_2 that is slightly lower than the refractive index of the core region. The cladding of the fiber is made of a high-silica-content glass or a multicomponent glass. Figure 8.9 shows the dimensions and refractive indexes of commonly used telecommunication fibers. Figure 8.10 enumerates some of the advantages, constraints, and applications of the different types of fibers. In general, when the transmission medium must have a very high bandwidth—for example, in an undersea or long-distance terrestrial system—a single-mode fiber is used. For intermediate system bandwidth requirements between 150 MHz-km and 2 GHz-km, such as found in local-area networks, either a single-mode or graded-index multimode fiber would be the choice. For applications such as short data links in which lower bandwidth requirements are placed on the transmission medium, either a graded-index or a step-index multimode fiber can be used.

Because of their low loss and high-bandwidth capabilities, optical fibers have the potential for being used wherever twisted-wire pairs or coaxial cables are used as the transmission medium in a communication system. If an engineer were interested in choosing a transmission medium for a given transmission objective, he or she would tabulate the required and desired features of alternative technologies that may be available for use in the applications. With that process in mind, a summary of the attractive features and the advantages of optical fiber transmission will be given. These advantages include (1) low loss

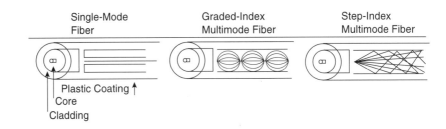

FIGURE 8.10 Advantages, constraints, and applications of optical fibers.

and high bandwidth; (2) small size and bending radius; (3) nonconductive, nonradiative, and noninductive; (4) light weight; and (5) providing natural growth capability.

To appreciate the low- and high-bandwidth capabilities of optical fibers, consider the curves of signal attenuation vs frequency for three different transmission media shown in Figure 8.11. Optical fibers have a "flat" transfer function well beyond 100 MHz. Compared with wire pairs or coaxial cables, optical fibers have far less loss for signal frequencies above a few megahertz. This is an important characteristic that strongly influences system economics, because it allows the system designer to increase the distance between regenerators (amplifiers) in a communication system.

The small size, small bending radius (a few centimeters), and light weight of optical fibers and cables are very important where space is at a premium, such as in aircraft, on ships, and in crowded ducts under city streets.

Because optical fibers are dielectric waveguides, they avoid many noise problems such as radiated interference, ground loops, and, when installed in a cable without metal, lightning-induced damage that exists in other transmission media.

Finally, the engineer using optical fibers has a great deal of flexibility. He or she can install an optical fiber cable and use it initially in a low-capacity (low-bit-rate) system. As the system needs to grow, the engineer can take advantage of the broadband capabilities of optical fibers and convert to a high-capacity (high-bit-rate) system by simply changing the terminal electronics.

The proper design and operation of an optical communication system using optical fibers as the transmission medium require a knowledge of the transmission characteristics of the optical sources, fibers, and interconnection devices (connectors, couplers, and splices) used to join lengths of fibers together. The transmission criteria that affect the choice of the fiber type used in a system are

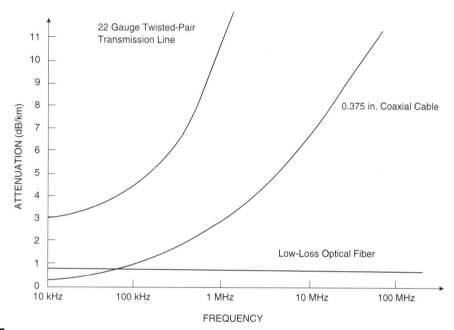

FIGURE 8.11 Signal attenuation vs frequency for different transmission media.

TABLE 8.1 Loss mechanisms in optical fibers

1. Intrinsic material absorption loss
   (a) Ultraviolet absorption tail
   (b) Infrared absorption tail
2. Absorption loss due to impurity ions
3. Rayleigh scattering loss
4. Waveguide scattering loss
5. Microbending loss

signal attenuation, information transmission capacity (bandwidth), and source coupling and interconnection efficiency. Signal attenuation is due to a number of loss mechanisms within the fiber, as shown in Table 8.1, and due to the losses occurring in splices and connectors. The information transmission capacity of a fiber is limited by dispersion, a phenomenon that causes light that is originally concentrated into a short pulse to spread out into a broader pulse as it travels along an optical fiber. Source and interconnection efficiency depends on the fiber's core diameter and its numerical aperture, a measure of the angle over which light is accepted in the fiber. Absorption and scattering of light traveling through a fiber lead to signal attenuation, the rate of which is measured in decibels per kilometer (dB/km). As can be seen in Figure 8.12, for both multimode and single-mode fibers, attenuation depends strongly on wavelength. The decrease in scattering losses with increasing wavelength is offset by an increase in material absorption such that attenuation is lowest near $1.55\ \mu m$ (1550 nm).

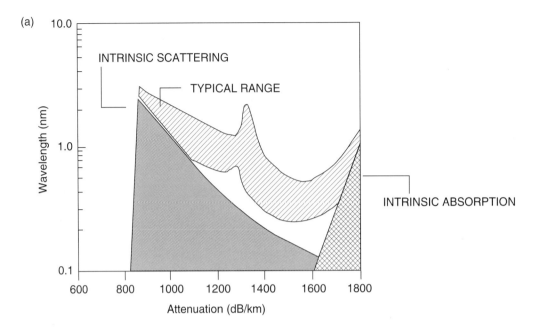

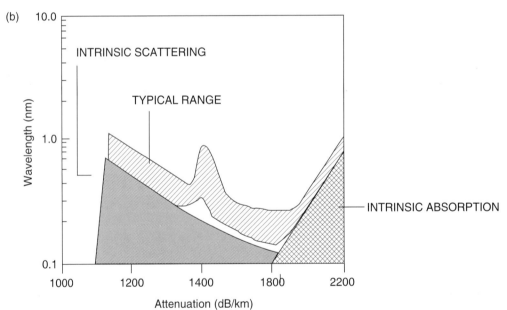

FIGURE 8.12 Dependence of attenuation on wavelength (a) for multimode and (b) single-mode fibers.

The measured values given in Table 8.2 are probably close to the lower bounds for the attenuation of optical fibers. In addition to intrinisic fiber losses, extrinsic loss mechanisms, such as absorption due to impurity ions and microbending loss due to jacketing and cabling, can add loss to a fiber. The bandwidth or information-carrying capacity of a fiber is inversely related to

TABLE 8.2 Best attenuation results (dB/km) in Ge-P-SiO$_2$ core fibers

Wavelength (nm)	$\Delta \approx 0.2\%$ (single-mode fiber)	$\Delta \approx 1.0\%$ (graded-index multimode fiber)
850	2.1	2.20
1300	0.27	0.44
1500	0.16	0.23

its dispersion. The total dispersion in a fiber is a combination of three components: intermodal dispersion (modal delay distortion), material dispersion, and waveguide dispersion.

Intermodal dispersion occurs in multimode fibers because rays associated with different modes travel different effective distances through the optical fiber. This causes light in the different modes to spread out temporally as it travels along the fiber. Modal delay distortion can severely limit the bandwidth of a step-index multimode fiber to the order of 20 MHz/km. To reduce modal delay distortion in multimode fibers, the core is carefully doped to create a graded (approximately parabolic) refractive index profile. By carefully designing this index profile, the group velocities of the propagating modes are nearly equalized. Bandwidths of 1.0 GHz-km are readily attainable in commercially available graded-index multimode fibers. The most effective way to eliminate intermodal dispersion is to use a single-mode fiber. As only one mode propagates in a single-mode fiber, modal delay distortion between modes does not exist and very high bandwidths are possible. The bandwidth of a single-mode fiber, as mentioned previously, is limited by the combination of material and waveguide dispersion. As shown in Figure 8.13, both material and waveguide dispersions are dependent on wavelength.

Material dispersion is caused by the variation of the refractive index of the glass with wavelength and the spectral width of the system source. Waveguide dispersion occurs because light travels in both the core and cladding of a single-mode fiber at an effective velocity between that of the core and cladding materials. The waveguide dispersion arises because the effective velocity changes with wavelength. The amount of waveguide dispersion depends on the design of the waveguide structure as well as on the fiber material. Material and waveguide dispersions are measured in picoseconds (of pulse spreading) per nanometer (of source spectral width) per kilometer (of fiber length), reflecting both the increases in magnitude in source linewidth and the increase in dispersion with fiber length.

Material and waveguide dispersions can have different signs and effectively cancel each other's dispersive effect on the total dispersion in a single-mode fiber. In conventional germanium-doped silica fibers, the "zero-dispersion" wavelength at which the waveguide and material dispersion effects cancel each other out occurs near 1.30 μm. The zero-dispersion wavelength can be shifted to 1.55 μm, or the low-dispersion characteristics of a fiber can be broadened

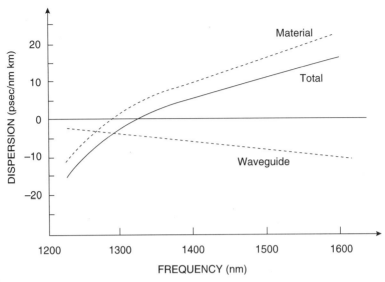
FIGURE 8.13 Dependence of waveguide dispersion on wavelength.

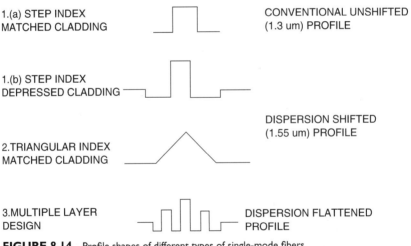
FIGURE 8.14 Profile shapes of different types of single-mode fibers.

by modifying the refractive index profile shape of a single-mode fiber. This profile shape modification alters the waveguide dispersion characteristics of the fiber and changes the wavelength region in which waveguide and material dispersion effects cancel each other. Figure 8.14 illustrates the profile shapes of conventional, dispersion-shifted, and dispersion-flattened single-mode fibers. Single-mode fibers operating in their zero-dispersion region with system sources of finite spectral width do not have infinite bandwidth (and may exhibit polarization mode dispersion) but have bandwidths that are high enough to satisfy all current high-capacity system requirements.

8.3. ANALYSIS OF OPTICAL FIBERS

8.3.1. The Step-Index Fiber

In this section electromagnetic field theory will be used to solve rigorously the boundary value problem of the round optical fiber with a homogeneous core (the step-index fiber) (Morse and Feshbach 1953). To simplify this analysis, we will assume that the fiber is oriented along the z axis and the radius b of the fiber cladding is large enough to ensure that the cladding field decays exponentially and approaches zero at the cladding–air interface. This allows us, as shown in Figure 8.15, to analyze the fiber as a tractable two-media boundary value problem. This thick-cladding assumption agrees well with the conditions that exist within a properly designed optical fiber used for communication purposes. The steps that we will follow to solve the boundary value problem of the step-index fiber are outlined in Table 8.3. To obtain the modes in a step-index optical fiber, one must solve the modified wave equation [Eqs. (8.1) and (8.2)] shown

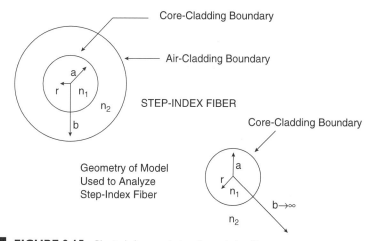

FIGURE 8.15 Physical characteristics of step-index fibers.

TABLE 8.3 Analysis of the step-index fiber procedures followed

1. Mathematically model the step-index fiber using the wave equation in cylindrical coordinates.
2. Use the technique of separation of variables to partition the wave equation.
3. Define the physical requirements that influence the solution of the fields in the core and cladding.
4. Select the proper functional form of the solution of the modified wave equation (Bessel's equation) in the core and cladding.
5. Apply the boundary conditions at the core–cladding interface.
6. Obtain the "characteristic" equation and its resulting modal equations.
7. Analyze the resulting modes and their cutoff conditions.

later for E_z and H_z in both the core and cladding regions of the fiber.

$$\frac{\partial^2 E_z}{\partial r^2} + \frac{1}{r}\frac{\partial E_z}{\partial r} + \frac{1}{r^2}\frac{\partial^2 E_z}{\partial \phi^2} + \kappa^2 E_z = 0 \qquad (8.1)$$

$$\frac{\partial^2 H_z}{\partial r^2} + \frac{1}{r}\frac{\partial H_z}{\partial r} + \frac{1}{r^2}\frac{\partial^2 H_z}{\partial \phi^2} + \kappa^2 H_z = 0 \qquad (8.2)$$

Having obtained expressions for E_z and H_z, we can directly obtain from Maxwell's equations expressions for the transverse components of the fields.

$$E_r = -\frac{j}{\kappa^2}\left(\beta\frac{\partial E_z}{\partial r} + \omega\mu\frac{1}{r}\frac{\partial H_z}{\partial \phi}\right) \qquad (8.3)$$

$$E_\phi = -\frac{j}{\kappa^2}\left(\beta\frac{1}{r}\frac{\partial E_z}{\partial \phi} - \omega\mu\frac{\partial H_z}{\partial r}\right) \qquad (8.4)$$

$$H_r = -\frac{j}{\kappa^2}\left(\beta\frac{\partial H_z}{\partial r} - \omega\varepsilon\frac{1}{r}\frac{\partial E_z}{\partial \phi}\right) \qquad (8.5)$$

$$H_\phi = -\frac{j}{\kappa^2}\left(\beta\frac{1}{r}\frac{\partial H_z}{\partial \phi} + \omega\varepsilon\frac{\partial E_z}{\partial r}\right) \qquad (8.6)$$

Because Eqs. (8.1) and (8.2) have the same mathematical form, we will solve Eq. (8.1), understanding that solutions obtained for it will be valid for Eq. (8.2). To obtain Eq. (8.1) we have already assumed an optical system with cylindrical symmetry. The longitudinal direction of propagation is the z axis and the dependence of the fields is of the form $e^{j(\omega t - \beta z)}$.

The technique of separation of variables will now be applied to obtain a solution of Eq. (8.1). We will assume that we can obtain independent solutions for E_z in ϕ and r, that is,

$$E_z(\phi, r) = A\Phi(\phi)F(r) \qquad (8.7)$$

Since the fiber has circular symmetry, we will choose a circular function as a trial solution for $\Phi(\phi)$.

$$\Phi(\phi) = e^{j\upsilon\phi} \qquad (8.8)$$

where υ is a positive or negative integer. Now we have

$$E_z AF(r)e^{j\upsilon\phi} \qquad (8.9)$$

Taking the second-order derivatives of Eq. (8.9) with respect to r and ϕ and substituting back into Eq. (8.1), we obtain

$$\frac{d^2 F(r)}{dr^2} + \frac{1}{r}\frac{dF(r)}{dr} + \left(\kappa^2 - \frac{\upsilon^2}{r^2}\right)F(r) = 0 \qquad (8.10)$$

Equation (8.10) is a form of Bessel's equation where κ is defined as the wave number and given by the expression $\kappa^2 = k^2 - \beta^2$, k being the complex propagation constant equal to $\omega^2\mu\varepsilon$. This well-known second-order differential equation has two independent solutions. Numerous cylinder functions satisfy

ANALYSIS OF OPTICAL FIBERS

Bessel's equation. Energy considerations will dictate the choice of the functions selected as solutions of Eq. (8.10); that is,

1. The field must be finite in the core of the fiber. Specifically the cylinder function chosen in the core of the fiber must be finite at $r = 0$.
2. The field in the cladding of the fiber must have an exponentially decaying behavior at large distances from the center of the fiber.

Because the fields must be finite at the center of the fiber core, we will choose $J_\upsilon(\kappa r)$ as the form of the solution for $r < a$. Therefore, for $r < a$,

$$E_z = A J_\upsilon(\kappa r) e^{j\upsilon\phi} \tag{8.11}$$

$$H_z = B J_\upsilon(\kappa r) e^{j\upsilon\phi} \tag{8.12}$$

We require that the field in the cladding of the fiber decay in the r direction and be of the form $e^{-\gamma r}$ where γ is the decay constant of the evanescent field, given by $\gamma^2 = \beta^2 - n_2^2 k_0^2$, with n_2 being the index of refraction of the cladding given by $(\varepsilon_{r_2})^{1/2}$.

If we define $\kappa = j\gamma$ we can choose a modified Hankel function of the first kind to describe the decaying behavior of the field in the cladding for large r. That is, for $r > a$,

$$E_z = C H_\upsilon^{(1)}(j\gamma r) e^{j\upsilon\phi} \tag{8.13}$$

$$H_z = D H_\upsilon^{(1)}(j\gamma r) e^{j\upsilon\phi} \tag{8.14}$$

where A, B, C, and D are unknown constants to be determined.

To obtain the transverse fields in the core and cladding of the guide, one must use Eqs. (8.3) through (8.6). For example, to obtain E_r in Eq. (8.3) for both $r < a$ and $r > a$ one must differentiate the longitudinal fields (i.e., E_z) of Eqs. (8.11) and (8.13), respectively, with respect to r and ϕ and then substitute the results back into Eq. (8.3). After some simplification we obtain, for $r < a$,

$$E_r = -\frac{j}{\kappa^2}\left[A\beta\kappa J_\upsilon'(\kappa r)e^{j\upsilon\phi} + B(j\upsilon)(\omega\mu)\frac{1}{r}J_\upsilon(\kappa r)e^{j\upsilon\phi}\right]e^{j\upsilon\phi} \tag{8.15}$$

For $r > a$ we have

$$E_r = -\frac{1}{\gamma^2}\left[\beta\gamma C H_\upsilon^{(1)\prime}(j\gamma r) + \omega\mu_0 \frac{\upsilon}{r} D H_\upsilon^{(1)}(j\gamma r)\right]e^{j\upsilon\phi} \tag{8.16}$$

where primed terms (i.e., terms with ′) mean the first derivatives with respect to κr. Furthermore, for the core region ($r < a$)

$$\kappa^2 = k_1^2 - \beta^2 \tag{8.17}$$

$$k_1^2 = \omega^2 \mu_0 \varepsilon_1 \tag{8.18}$$

and for the cladding region ($r > a$)

$$\gamma^2 = \beta^2 - k_2^2 \tag{8.19}$$

$$k_2^2 = \omega^2 \mu_0 \varepsilon_2 \tag{8.20}$$

In a similar manner using Eqs. (8.4) through (8.6), we can obtain for the core region ($r > a$)

$$E_\phi = -\frac{j}{\kappa^2}\left[j\beta\frac{\upsilon}{r}AJ_\upsilon(\kappa r) - \kappa\omega\mu BJ'_\upsilon(\kappa r)\right]e^{j\upsilon\phi} \qquad (8.21)$$

$$H_r = -\frac{j}{\kappa^2}\left[-j\omega\varepsilon_1\frac{\upsilon}{r}AJ_\upsilon(\kappa r) + \kappa\beta BJ'_\upsilon(\kappa r)\right]e^{j\upsilon\phi} \qquad (8.22)$$

$$H_\phi = -\frac{j}{\kappa^2}\left[\kappa\omega\varepsilon_1 AJ'_\upsilon(\kappa r) + j\beta\frac{\upsilon}{r}BJ_\upsilon(\kappa r)\right]e^{j\upsilon\phi} \qquad (8.23)$$

and for the cladding region ($r > a$)

$$E_\phi = -\frac{1}{\gamma^2}\left[\beta\frac{\upsilon}{r}CH^{(1)}_\upsilon(j\gamma r) - \gamma\omega\mu_0 DH^{(1)'}_\upsilon(j\gamma r)\right]e^{j\upsilon\phi} \qquad (8.24)$$

$$H_r = -\frac{1}{\gamma^2}\left[-\omega\varepsilon_2\frac{\upsilon}{r}CH^{(1)}_\upsilon(j\gamma r) - \gamma\beta DH^{(1)'}_\upsilon(j\gamma r)\right]e^{j\upsilon\phi} \qquad (8.25)$$

$$H_\phi = -\frac{1}{\gamma^2}\left[\gamma\omega\varepsilon_2 CH^{(1)'}_\upsilon(j\gamma r) + \beta\frac{\upsilon}{r}DH^{(1)}_\upsilon(j\gamma r)\right]e^{j\upsilon\phi} \qquad (8.26)$$

The constants A, B, C, D, and β are determined by applying the boundary conditions for the two tangential components of the electric and magnetic fields at the core–cladding interface ($r = a$). The boundary conditions for the fields at the core–cladding interface can be written as

$$E_{z_1} = E_{z_2}$$
$$E_{\phi_1} = E_{\phi_2} \qquad \text{for } r = a$$
$$H_{z_1} = H_{z_2}$$
$$H_{\phi_1} = H_{\phi_2}$$

where subscripts 1 and 2 refer to the fields in the core and cladding, respectively. Applying these conditions yields four simultaneous equations for the unknown A, B, C, and D. This solution yields a determinant. The solutions for A, B, C, and D can be obtained from this determinant provided that the system determinant for the four equations vanishes. Expansion of this determinant results in what is known as the "eigenvalue" or characteristic equation of the waveguide. This equation defines the modes in the guide and yields the permissible values of β, κ, and γ associated with each mode. The resulting characteristic equation for the setp-index fiber is

$$\left[\frac{\varepsilon_1}{\varepsilon_2}\frac{a\gamma^2}{\kappa}\frac{J'_\upsilon(\kappa a)}{J_\upsilon(\kappa a)} + j\gamma a\frac{H^{(1)'}_\upsilon(j\gamma a)}{H^{(1)}_\upsilon(j\gamma a)}\right] \times \left[\frac{a\gamma^2 J'_\upsilon(\kappa a)}{\kappa J_\upsilon(\kappa a)} + j\gamma a\frac{H^{(1)'}_\upsilon(j\gamma a)}{H^{(1)}_\upsilon(j\gamma a)}\right]$$

$$= \left[\upsilon\left(\frac{\varepsilon_1}{\varepsilon_2} - 1\right)\frac{\beta k_2}{\kappa^2}\right]^2 \qquad (8.27)$$

The coefficients A, B, C, and D can be written so that A is the only unknown coefficient. For example, it can be shown that

$$C = \frac{J_\upsilon(\kappa a)}{H_\upsilon^{(1)}(j\gamma a)} A \tag{8.28}$$

$$D = \frac{J_\upsilon(\kappa a)}{H_\upsilon^{(1)}(j\gamma a)} B \tag{8.29}$$

where A and B are related to each other via

$$B = j\upsilon \frac{\omega(\varepsilon_1 - \varepsilon_2)\beta J_\upsilon(\kappa a) H_\upsilon^{(1)}(j\gamma a)}{\kappa\gamma a \left[\gamma J_\upsilon'(\kappa a) H_\upsilon^{(1)}(j\gamma a) + j\kappa J_\upsilon(\kappa a) H_\upsilon^{(1)'}(j\gamma a)\right]} A \tag{8.30}$$

In general, the permissible field configurations or modes that exist in a step-index fiber have six field components. For the round fiber, hybrid modes exist as well as the transverse electric (TE) and transverse magnetic (TM) modes. The hybrid modes are denoted HE and EH modes and have both longitudinal electric and magnetic field components present. In terms of a ray analogy for the step-index fiber, the hybrid modes correspond to propagating skew rays and the TE and TM modes correspond to propagating meridional rays [$\upsilon = 0$ in Eq. (8.27)]. For meridional rays the right-hand side of Eq. (8.27) is equal to zero and one obtains two characteristic equations that define the TE and TM modes. These equations are

$$\left[\frac{a\gamma^2}{\kappa}\frac{J_0'(\kappa a)}{J_0(\kappa a)} + j\gamma a \frac{H_0^{(1)'}(j\gamma a)}{H_0^{(1)}(j\gamma a)}\right] = 0 \tag{8.31}$$

$$\left[\frac{\varepsilon_1}{\varepsilon_2}\frac{a\gamma^2}{\kappa}\frac{J_0'(\kappa a)}{(\kappa a)} + \frac{j\gamma a H_0^{(1)'}(j\gamma a)}{H_0^{(1)}(j\gamma a)}\right] = 0 \tag{8.32}$$

An important parameter for each propagating mode is its cutoff frequency. A mode is cut off when its field in the cladding ceases to be evanescent and is detached from the guide; that is, the field in the cladding does not decay. The rate of decay of the fields in the cladding is determined by the value of the constant γ. For large values of γ, the fields are tightly concentrated inside and close to the core. With decreasing values of γ, the fields reach farther out into the cladding. Finally, for $\gamma = 0$, the fields detach themselves from the guide. The frequency at which this happens is called the cutoff frequency. The mode cutoff frequency can be calculated from Eqs. (8.17) through (8.20) for $\gamma = 0$ and is given by

$$\omega_c = \frac{\kappa_c}{\sqrt{\mu_0(\varepsilon_1 - \varepsilon_2)}} \tag{8.33}$$

The cutoff of each mode is obtained by solving the characteristic equation for κ_c for that mode. The cutoff frequency of a mode can be zero if $\kappa_c = 0$. One and only one mode can exist in an optical fiber with $\omega_c = 0$. This mode is the hybrid HE_{11} mode and it exists for all frequencies. If the fiber is designed to operate with only the HE_{11} mode present, it will operate as a single-mode optical fiber. The single-mode fiber has a very small core diameter and small refractive index difference between its core and cladding. These parameters are chosen to ensure that all other guided modes are below their cutoff frequency. To better understand the relationship between mode cutoff and the physical parameters of a fiber, a cutoff parameter $\kappa_c a$, which is usually called the "V" number of the fiber, is defined as

$$\kappa_c a = \omega_c \sqrt{\mu_0 \varepsilon_0} \left(\sqrt{n_1^2 - n_2^2} \right) a \tag{8.34}$$

Noting that

$$\omega_c \sqrt{\mu_0 \varepsilon_0} = \frac{2\pi}{\lambda_0} \tag{8.35}$$

we finally obtain

$$V \equiv \kappa_c a = \frac{2\pi a}{\lambda_0} \sqrt{n_1^2 - n_2^2} \tag{8.36}$$

The number of propagating modes in a step-index fiber is proportional to the V number as shown in Figure 8.16. Table 8.4 illustrates how increasing a, n_1, n_2, or λ_0 influences the number of propagating modes in the fiber. Notice that for $V < 2.405$ it is possible to design a single-mode fiber that supports only the HE_{11} mode. Typical commercial single-mode fibers have a core diameter of approximately 8–10 µm and a refractive index difference between the core and the cladding of approximately 0.2%. These types of fibers are used for long-distance communications where their very large information-carrying capacity is needed. Single-mode fibers are used in a large majority of the fiber-optic telecommunication systems being installed today. A typical commercial step-index multimode fiber, on the other hand, will propagate hundreds of mode

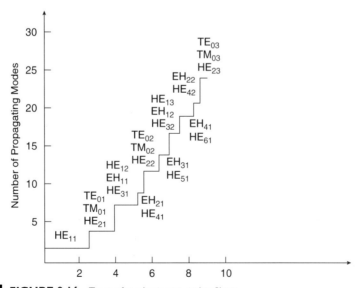

FIGURE 8.16 Types of modes in step-index fibers.

TABLE 8.4 Dependence of the number of propagating modes in a fiber on the physical properties of the fiber

Increasing physical parameters	Number of propagating modes
Core radius, a	Increases
Core refractive index, n_1	Increases
Cladding refractive index, n_2	Decreases
Source wavelength, λ_0	Decreases

groups and have a relatively large core diameter and refractive index difference between the core and cladding. Step-index multimode fibers are used in short-distance data links where their lower information-carrying capacity is not an issue.

8.4. THE GRADED-INDEX FIBER

The graded-index fiber, because of its relatively large bandwidth and core diameter, is used in many local-area networks where moderate information-carrying capacity is needed. The multimode step-index fiber bandwidth is severely limited (less than 100 MHz-km) due to modal delay distortion. The grading of the refractive index profile of a fiber core has the effect of increasing the bandwidth of a fiber to up to 2 GHz-km by equalizing the group delays of the various propagating mode groups.

Let's consider a multimode fiber with an inhomogeneous core as shown in Figure 8.17. The wave Eq. (8.10) is rewritten below showing the variation of the refractive index with r:

$$\frac{d^2 F(r)}{dr^2} + \frac{1}{r}\frac{dF(r)}{dr} + \left[k^2(r) - \beta^2 - \frac{v^2}{r^2}\right]F(r) = 0 \qquad (8.37)$$

where

$$k(r) = \frac{2\pi}{\lambda_0}n(r) = k_0 n(r) \qquad (8.38)$$

and

$$\kappa^2 = k^2(r) - \beta^2 \qquad (8.39)$$

The solution of the wave equation leads to a "characteristic equation" for the guide which relates β to κ. To simplify the analysis we first assume that the refractive index continues to decrease in the cladding following the core profile shown by the dashed curve in Figure 8.17 (i.e., refractive index is circular symmetric).

Furthermore, to solve the wave equation we resort to a ray optics method of analysis based on the WKBJ method (after Wentzel, Kramers, Brillouin, and Jefferies [Morse and Feshbach 1953]). The WKJB approach is a geometric optics approximation that works whenever the refractive index of the fibers

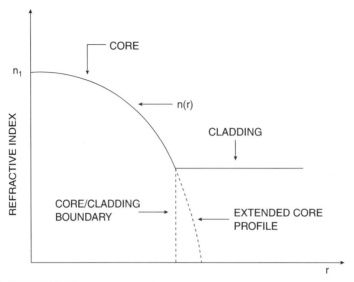

FIGURE 8.17 Physical profile of a graded-index fiber.

varies only slightly over distances on the order of the optical wavelength. Implementation of the WKBJ approach to a graded-index fiber yields that for a propagation mode to exist it is a necessary condition that (Cherin 1983)

$$k^2(r) - \beta^2 - \frac{v^2}{r^2} > 0 \tag{8.40}$$

Figure 8.18a illustrates $k^2(r)$ and v^2/r^2 as a function of the radius r. The solid curve in Figure 8.18b shows $k^2(r) - v^2/k^2$ as a function of r. For a fixed value of β there exist two values of r (r_1 and r_2) which

$$k^2(r) - \frac{v^2}{r^2} - \beta^2 = 0 \tag{8.41}$$

It is between these two radii that the ray associated with the assumed plane-wave solution is constrained to move. Outside these two values of r, called caustics, the ray becomes imaginary, leading to decaying fields.

For a fixed value of β, as v increases, the region between the two caustics becomes narrower. As v is increased further a point will be reached where the caustics merge. Beyond this point the wave is no longer bound. The propagation conditions of a wave depend on the value of both β and v. For a fixed value of v, modes far from cutoff have large β values and correspondingly more closely spaced caustics. In general, a bound hybrid mode in a graded-index optical fiber can be represented pictorially by a skew ray spiraling down the fiber between two caustics. Both inside ($r < r_1$) and outside ($r < r_2$) the caustics, the field corresponding to the hybrid mode decays.

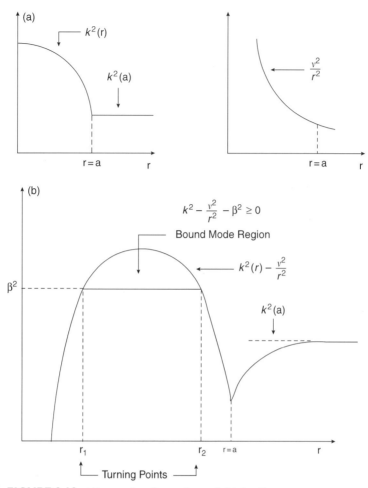
FIGURE 8.18 Wave number diagram for graded-index fiber.

8.5. CT SCANNERS IN MEDICINE

For almost a century, x rays have been used for medical imaging and radiation therapy. Over 100 years ago Wilhelm Röntgen, a professor of physics at the Julius Maximilian University of Wurzburg, discovered x rays while experimenting with cathode rays in a Crookes tube. Word of this discovery spread quickly, and by early 1986 the properties of x rays were under investigation in many laboratories in Europe and North America. By the turn of the century, physicians were exploiting the penetrating character of x rays to look inside the human body without cutting it open.

The usage of x rays for medical diagnosis and therapy has expanded enormously since those early years. Today, in the United States alone, over 300 million clinical x-ray examinations are performed annually for purposes ranging from static imaging of fractures and cancers to the real-time guidance of tissue biopsies and cardiovascular angioplasties. In addition, half a million cancer patients each year receive x-ray treatments, about half of them for curative purposes and the rest for pain relief.

Until recently the diagnostic and therapeutic applications were distinct. Today, however, the boundary between the diagnostic and therapeutic applications of x rays in medicine is far less distinct.

Ordinary planar x-ray images are formed by placing a patient between an x-ray tube and an image receptor, usually a cassette containing an intensifying screen and a photographic film. The film is exposed by light emitted when the transmitted x rays interact in the screen. The resulting radiograph is a static shadow image. Fluoroscopy is a variant of this procedure in which a fluorescent screen and an electronic image intensifier are used to form a continuous moving picture. When the x rays traverse the patient, they can be absorbed, scattered or transmitted undisturbed to the receptor. The scattered x rays merely interfere with the information conveyed by the shadow pattern of transmitted rays. Therefore, a mechanical grid is inserted behind the patient to prevent most of the scattered x rays from reaching the cassette. For a parallel, monoenergetic x-ray beam incident along the z axis, the distribution $N(x, y)$ of transmitted x-ray photons at the image plane is given, in the absence of scattering, by

$$N_o A \int e^{-\mu(z)} dz$$

where the line integral is taken over all tissues along the unscattered photon trajectory to the point (x, y) on the image plane, μ is the linear attenuation coefficient for x rays of the tissue encountered at (x, y, z), and A is the x-ray energy absorption coefficient of the intensifying screen. The distribution of x rays absorbed in the screen thus forms a two-dimensional projection image of the transmission of x rays through the three-dimensional volume of tissue exposed to the x-ray beam.

The linear attenuation coefficient μ is in fact the sum of the coefficients for various types of x-ray interactions. For the range of x-ray energies employed in medical imaging, two kinds of interactions predominate: the photoelectric effect, described by the linear attenuation coefficient τ, and Compton scattering described by the linear attenuation coefficient σ. Thus $\mu = \tau + \sigma$. Figure 8.19 shows the x-ray energy dependencies of these coefficients in human soft tissue. The photoelectric coefficient increases with atomic number Z like Z^3, principally because x rays interact photoelectrically with the inner, tightly bound electrons of an atom. The Compton coefficient, by contrast, is relatively independent of the atomic number of the tissue atoms, because x rays Compton scatter almost exclusively off the outer, loosely bound atomic electrons. Both coefficients increase linearly with the tissue density. To achieve contrast between soft tissues that differ only slightly in Z, one must use low-energy x rays, because they interact predominantly by the photoelectric effect. An example is in mammography, which employs x rays in the range of 15–30 keV. For chest x-ray images, which involve tissue of greater intrinsic contrast, clinicians use x rays with energies ranging from 50 to 150 keV.

X-ray images represent a combination of four kinds of resolution: spatial, contrast, temporal, and statistical. Improving any one of these resolution

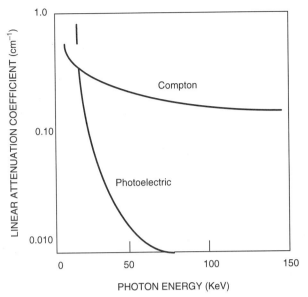

FIGURE 8.19 Linear attenuation coefficient for x rays traversing human soft tissue is the sum of two dominating contributions.

factors degrades one or more of the others. The compromise among them represents a balance such that no single factor dominates the degradation of the image.

8.5.1. Sectional Imaging

A limitation of conventional planar x-ray imaging is the projection of a three-dimensional distribution of attenuation coefficients as a shadow onto a two-dimensional detector. This type of projection discards a lot of information about tissue variation along the beam direction. For many years, techniques of analog tomography were used in attempts to overcome this limitation, but they were restricted to certain applications, and images were difficult to interpret.

An important development was achieved in 1972 with the introduction of x-ray transmission computed tomography (CAT). This technique was brought to clinical medicine through the effort of Godfrey Hounsfield and Allan Cormack, who shared the 1979 Nobel Prize in medicine.

Consider a highly collimated x-ray pencil beam in the plane of a slice of the body only a few millimeters thick. X rays transmitted all the way through the slice are measured with a collimated detector on the opposite side of the patient. The signal from the x-ray detector is converted to digital output. The tight collimation of source and detector prevents scattered radiation from degrading image contrast. The number of x-ray photons recorded by the detector at one position constitutes a single pencil-beam projection of x-ray transmission data at a specific angle through the tissue slice. This process is repeated many times at slightly different angles to create a set of multiple projections of the entire tissue slice (Figure 8.20).

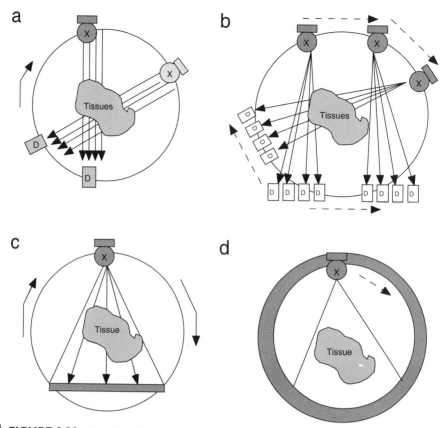

FIGURE 8.20 Evolution of geometries of x-ray CT scanners.

If the x-ray projection data are collected at a sufficient number of angles, a matrix of values of the attenuation coefficient μ for different $\delta x \delta y$ cells can be calculated by a simple back-projection technique, which yields the two-dimensional distribution $\mu(x, y)$ over the whole tissue slice. In displaying the variation of the attenuation coefficient pictorially in shades of gray, one creates an image that shows the various anatomical features of the tissue slice. In practice, the back projection is calculated by Fourier transforming the projection data into (spatial) frequency space.

The first type of CT scanners used a single collimated x-ray source and two detectors for data to be collected from two contiguous tissue slices. The source and each detector mapped out projection data in a translate-rotate geometry, one x-ray path at a time. The efforts to collect the data faster soon led to the successive generation of fan-beam scanners shown in Figures 8.20b, 8.20c, and 8.20d. Figure 8.20b shows an array of detectors that move in a translate-rotate configuration; Figure 8.20c shows a bank of detectors that move in a purely rotational geometry; finally Figure 8.20d shows a ring array of stationary detectors. Using fan-beam geometries, a whole tissue slice can be imaged in a few seconds. By combining these geometries with the patient gantry moving continuously along its long axis, one can get cross-sectional images of many slices of the patient in minimum time; a procedure called spiral scanning.

Although the evolution of x-ray scanning geometries greatly shortened the time required to acquire images, none of the three fan-beam generations permitted data acquisition in a time (less than 0.1 s) short enough to capture images of the heart and other blood-perfused organs without significant degradation caused by motion. Fifth-generation scanners employ an electron-beam gun that generates x-ray beams in different directions by scanning over a stationary concave metal target. The resulting scan times are only a few milliseconds.

8.5.2. Digital Imaging

The combination of intensifying screen and photographic film has many advantages for capturing and recording many x-ray images. This approach is simple, inexpensive, and it yields excellent spatial resolution. This approach is limited to a narrow range of acceptable exposures and offers little flexibility for image processing or data compression. Film images are bulky to store, and they must be transported from place to place. Digital imaging methods overcome these limitations, but they are more expensive and more complex. Digital methods employ a variety of approaches for x-ray detection and measurements: fluorescent crystals with photomultipliers, semiconductor detectors, channel electron multipliers, and photostimulatable phosphors with laser-scanning readout.

A digital radiographic unit (Figure 8.21), the x-ray source, and receptor are computer controlled to provide digital images that can be displayed in real time on video screens. Digital images can be stored on magnetic media. Digital image storage and display are used routinely in x-ray CT and magnetic resonance imaging.

8.6. THE ENDOSCOPE

The future of diagnostic devices using photonics looks promising with a large amount of research focusing on early cancer detection. Photonic components for the detection of cancer and other diseases keep getting smaller and cheaper, making many of these instruments fit inside the body. One such device, the endoscope, has been used for many years to diagnose and treat gastrointestinal problems, and research is still being performed to improve this instrument.

The endoscope uses a CCD or optical fibers to form images that are transmitted to a monitor. Endoscopes not only let doctors see inside, but also include an instrument that can take a biopsy, and some can even dye the area for x-ray imaging. By guiding an endoscope into the gastrointestinal tract, doctors can view lesions or sources of bleeding. Endoscopes for this purpose cost around $15,000 and typically have lifetimes of about 3000 uses. The endoscope can improve its image quality and get smaller to give the patient more comfort. Furthermore, the methods of cleaning and sterilizing the reusable instruments are complicated and time consuming. Attempts to make semidisposable endoscopes have not had great success. Patients sometimes avoid the uncomfortable procedure because it requires sedation and/or local anesthetic.

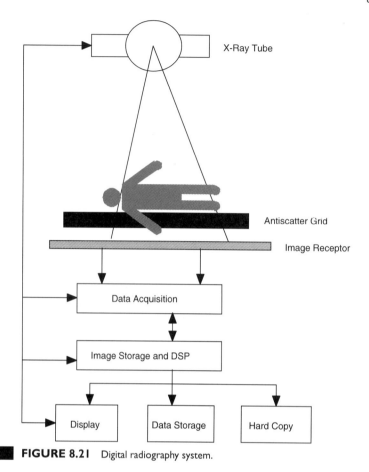

FIGURE 8.21 Digital radiography system.

In addition, endoscopes cannot reach all parts of the small intestine. However, making endoscopes that can go wireless would allow doctors to take endoscopes where they could not go before and could make patients more comfortable. An endoscope is illustrated in Figure 8.22.

8.7. DIGITAL X RAYS

Breast cancer remains a leading cause of cancer deaths in the world (50,000 + a year in the United States alone). Better diagnosis and treatment of breast cancer have noticeably improved the outcome of the disease, reducing death rates over the last decade by about 2% per year in the United States. A great deal of that success is due to earlier detection by the standard breast imaging technology, film-screen x-ray mammography.

The variety and sophistication of imaging technologies have increased greatly, encompassing everything from optical laser imaging to digital mammography. While these new technologies (see Table 8.5) have shown impressive

TABLE 8.5 Imaging technologies for breast cancer

Technology	Description	FDA approved	Used routinely	Used frequently	Clinical data suggest a role	Preclinical data suggest a role	Data not available
Film-screen mammography	The standard x-ray technique	Yes	Screening and diagnosis				
Digital mammography	Digital version of x-ray technique	Yes		Screening and diagnosis			
Ultrasound	Forms images by reflection of megahertz frequency sound waves	Yes	Diagnosis		Screening		
MRI	Forms images from radio emissions from nuclear spins	Yes		Diagnosis	Screening		
Scintimammography	Senses tumors from gamma ray emissions of radioactive pharmaceuticals	Yes			Diagnosis	Screening	
Thermography	Seeks tumors by infrared signature	Yes			Diagnosis	Screening	
Optical imaging	Localizes tumors by measuring scattered near infrared light				Diagnosis	Screening	
Electrical potential measurement	Identifies tumors by measuring potential at array of detectors on skin				Diagnosis	Screening	
Positron emission tomography	Forms images using emissions from annihilation of positrons from radioactive pharmaceuticals	Yes				Screening and diagnosis	
Novel ultrasound techniques	Include compound imaging, which improves resolution using sound waves from several angles; 3D and Doppler imaging					Screening and diagnosis	
Elastography	Uses ultrasound or MRI to infer the mechanical properties of tissue					Screening and diagnosis	
Magnetic resonance spectroscopy	Analyzes tissue's chemical makeup using radio emissions from nuclear spins					Screening and diagnosis	
Thermoacoustic computed tomography	Generates short sound pulses within breast using RF energy and constructs a 3D image from them						Screening and diagnosis
Microwave imaging	Views breast using scattered microwaves						Screening and diagnosis
Hall-effect imaging	Picks up sonic vibrations of charged particles exposed to a magnetic field						Screening and diagnosis
Magnetomammography	Senses magnetic contrast agents collected in tumors						Screening and diagnosis

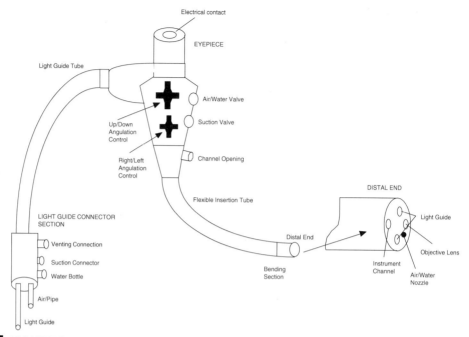
FIGURE 8.22 Flexible endoscope for looking at gastrointestinal tract without surgery.

results, a report released by the Institute of Medicine concludes that such technologies will only play a supporting role to film-screen mammography. Film mammography is the gold standard for screening for breast cancer and the technology against which all other technologies will be benchmarked.

Presently, abnormalities and lesions are discovered either by physical examination or by screening mammography, a task performed by a radiologist. Once identified, the abnormality must be diagnosed as benign or malignant by using other imaging technologies such as ultrasound or a biopsy and microscopic examination. The true tumors are biochemically characterized and categorized (staged) according to size and how much they have spread. The system is not flawless. It misses up to 20% of the tumors and many of those are found later to be benign.

Screening tools have to be highly sensitive, identifying as correctly as possible those tumors that could be malignant. Diagnostic tools must have a great specificity in order to really catch those tumors that are malignant.

Digital mammography (Figure 8.23) is the new technology most certain to see clinical use. With its high spatial resolution, mammography requires very small pixels and a high signal-to-noise ratio. The digital version of the technology has superior dynamic range and linearity compared to film, leading to a much better contrast resolution. It also allows the images to be manipulated and analyzed with software. This approach may lead to the discovery of more subtle features indicative of cancer and to a great ability to distinguish between potential cancers and harmless tissue abnormalities. Digital mammography may not make a difference in the numbers of cancers that are detected when compared

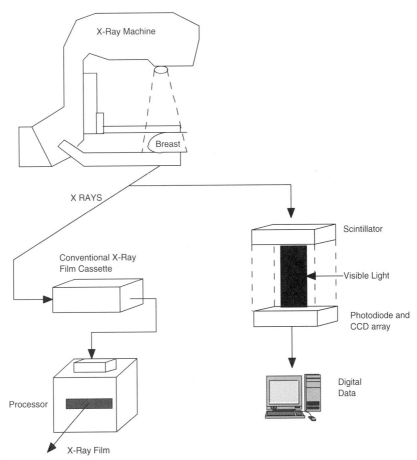

FIGURE 8.23 How digital mammography works.

with film mammography. However, it is also possible that digital mammography may decrease the number of follow-up tests needed for proper screening and diagnosis.

Digital mammography can employ either indirect detection, or scintillator x-ray conversion, or direct detection, or electronic x-ray conversion. In indirect detection, x rays are converted into visible light and then picked up on a solid-state detector. Direct detection, a more advanced technique aiming for higher resolution, converts x rays directly into electric signals. In the indirect detection, x rays pass through the breast and then strike a scintillator, a material that absorbs the x rays and emits visible light. The scintillator is coupled to a photodetector array or connected to tiles of (CCD) by tapered optical fibers. Cesium iodide can be used as a scintillator. An important design consideration in making indirect detectors is choosing the thickness of the scintillator. A thicker scintillator captures more x rays, but also leads to a loss of resolution because the photons have more chance to scatter before reaching the CCD. Direct detection omits the conversion of x rays to visible light, thus removing the loss from scattering of the light.

Clinical digital mammography provides digital files, rather than stacks of films. This allows the transfer of information more easily. Image enhancements and other software techniques will add enough functionalities for radiologists to make it worthwhile. Computer-aided detection (CAD) systems are software systems that can identify potential cancers on digital mammography images. These software systems are entering clinical practice as a way to improve radiologists' ability to detect the few cancer cases in the sea of normal-looking images they observe every day. In a typical mammography day, a radiologist uncovers around 6 cancers for every 1000 images observed. Some cancers are overlooked. Therefore, allowing the easy distribution of digital images among several radiologists can significantly reduce the chances of some cancers being missed. The use of CAD software has the peculiarity that the software has to be "trained" for a particular machine to recognize tumor sizes and types.

Other well-established imaging technologies such as MRI, positron emission tomography (PET), and ultrasound can assist digital mammography in the diagnosis problem. For example, physicians frequently use ultrasound to help determine whether a lesion detected on a mammogram is a malignant mass or a harmless cyst. However, many of these techniques have real limitations. MRI specificity is highly variable, ranging from 28 or 100 percent so, depending on the interpretation technique used and the patient population; PET scanners are expensive and scarce, and ultrasound has trouble seeing microcalcifications because of so-called speckle—tiny bright flecks on the image caused by scattered echoes.

The imaging principle relies on the physical characteristics of a cancer, such as its relative opacity to different wavelengths of light. This approach may need to be helped by combining the imaging approach with the usage of markers that exploit the biochemistry of cancer (e.g., compounds that will bind to tumors only) and that are picked up by imaging when such compounds react to light.

8.8. MEDICAL SENSORS FROM FIBER OPTICS

Fiber optics is beginning to have great use in healthcare for intracavity imaging and safe laser delivery. It is also becoming greatly useful in monitoring physiological functions. The optical fiber technology in healthcare provides many advantages.

1. The fibers are very small and flexible and they can be inserted inside very thin catheters and hypodermic needles. These are highly noninvasive techniques useful for monitoring.
2. Fibers are nontoxic, chemically inert, and very safe for patients. They are practically immune to electromagnetic interference from other electronic sources, which is of great importance for the patients.
3. Because of interference immunity there is no crosstalk between neighboring fibers, which allows the usage of several sensors grouped together in a single catheter.

The ideal fiber-optic sensors for medical applications must have the following properties: (1) reliability, (2) automated operations, and (3) simple implantation and low-cost maintenance. The fiber-optic sensors in medical applications are based on sensing processes that are typical of other types of fiber sensors, especially in intensity and time domain modulation. Mostly, they act as point sensors, either intrinsic or extrinsic, depending on whether the modulation is produced in the fiber sensor itself, or whether there is an external transducer connected to the fiber. They can also be used as spectral sensors, where the fibers act simply as light guiders.

The design of fiber-optic probes must be such that no thrombosis or inflammatory effects of the blood vessels occurs. Of course, this is only for temporary use since long-term implantable sensors are not feasible. Proper probe encapsulation is a crucial factor, so potential problems of selectivity, hysteresis, and stability must be resolved. Contrary to physical sensors, chemical and biochemical sensors cannot usually be hermetically encapsulated. There may be problems of interference with other substances in the environments.

8.8.1. Fiber Optics for Circulatory and Respiratory Systems

Fiber-optic sensors in the circulatory and respiratory systems are becoming very popular in both invasive and noninvasive approaches. Optical fibers can be used to measure the oxygen saturation in the blood. This measurement is necessary to monitor the cardiovascular and cardiopulmonary systems. The simplest method of this measurement is either through reflected or absorbed light, which is collected at two different wavelengths and the oxygen saturation is calculated on the basis of isosbestic regions of Hb (hemoglobin) and OxyHb (oxygenated hemoglobin) absorption.

Optical oximeters, which calculate oxygen saturation via the light transmitted to the earlobes, toes, and fingertips, have been developed primarily for neonatal care. A difficulty in neonatal care is the importance of differentiating between the light absorption due to arterial blood and that due to all other tissues and blood in the light path, which implies the use of multiple wavelengths. This challenge can be met using an oximeter. This approach is based on the assumption that a change in the light absorbed by the tissue during systole is caused by the passage of arterial blood. Using two wavelengths, it is possible to noninvasively measure the oxygen saturation by using the pulsatile rather than the absolute level of transmitted or reflected light intensity. The detection of blood absorbance fluctuations that are synchronous with systolic heart contractions is called photoplethysmography.

Blood Gases and Blood pH

Monitoring of the blood pH and blood oxygen (pO_2) and carbon dioxide (pCO_2) partial pressures to determine the quantity of oxygen delivered to the tissues and the quality of the perfusion is of great importance. The pH is detected by a chromophore, which changes its optical spectrum as a function of the pH. The absorption-based indicators or fluorophores are usually used. The carbon dioxide is detected indirectly, since its diffusion in a carbonate solution

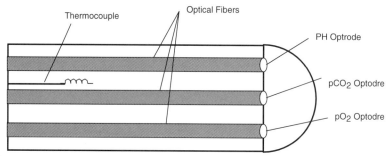

FIGURE 8.24 Fiber-optic vascular sensor for measuring blood oxygen and carbon dioxide.

fixed at the optical fiber tips alters the pH, which means that the CO_2 content can be determined by measuring the pH. Oxygen is measured using a separate chemical transducer; oxygen is detected via fluorescence techniques that exploit the quenching produced by oxygen on fluorophores.

The first intravascular sensor for simultaneous and continuous monitoring of the pH, pO_2, and pCO_2 is shown in Figure 8.24. It is composed of three optical fibers encapsulated in a polymer enclosure, including also a thermocouple used for temperature monitoring. Three fluorescent indicators are used as chromophores. The pH and pCO_2 are measured by the same fluorophore. The optoelectronics is composed of three modules, one for each sensor. A suitable filtered xenon lamp, when modulated, provides illumination for the modules. The source of light is focused through a lens onto a prism beam splitter and coupled through fiber to the sensor tip. The deflected light is collected by a reference detector for source control. The returning fluorescence is deflected by the prism beam splitter onto the signal detector.

There have been problems concerning the usage of intravascular fiber-optic sensors during clinical trials on volunteers in critical care and on surgical patients and these remain unresolved.

1. blood flow decreases due to peripheral vasoconstriction lasting for several hours after surgical operations, which can give rise to a contamination by flush solutions
2. the wall effect, which affects the oxygen count
3. the formation of a clot around the sensor tip, which can alter the value of all the analytical values.

Respiratory Monitoring

In intensive care units there is a need to continuously monitor breathing condition. It is possible to accomplish this monitoring from the nurse's station in such a way that patients can be kept under observation without the need for the nurses to be physically present. An optical fiber with a moisture-sensitive cladding can be developed for that purpose. The cladding is a plastic film doped with umbelliferous dye, which is a moisture-sensitive fluorescent material when pumped with UV light. The sensitive fiber section is placed over the patient's mouth and excited with He-Cd laser on a halogen lamp. Since the water vapor content in the human exhalation exceeds that in the room, the patient's exhalation produces a fluorescent signal that is detected by an electro-optical unit at the nurse's station. This monitoring is very useful in detecting abnormal breathing in patients.

Angiology

Blood vessels that are obstructed by atherosclerotic plaques can be recanalized by means of a pulse excimer laser radiation, guided by an optical fiber (laser angioplasty). Despite its widespread uses there is the possibility of blood vessel perforation, which occurs in between 20% and 40% of patients. To minimize the risk there is the usage of laser-induced fluoresce diagnosis of the vessel wall. This is part of identifying the target under irradiation. An all-optical approach to target identification is suggested by the fact that short optical pulses, when absorbed by the tissue, generate ultrasonic thermoelastic waves. The amplitude and temporal characteristics of the acoustic signal are dependent on the target composition and can be detected by a pressure fiber-optics system. In this approach an optical fiber tipped with a Fabry-Pérot cavity is inserted in the lumbar of the artery, and the sensor is in contact with the tissue. As shown in Figure 8.25, two signals are guided by the optical fiber: (1) pulse light of a Nd:YAG laser used to generate the thermoelastic wave in the tissue and (2) low-power continuous wave light of a tunable laser diode used for sensor interrogation. The Fabry-Pérot cavity of the fiber tip is formed by a polyethylene-terephthalate film and the fiber end provides a fiber–polymer acoustic impedance match. As the polymer film is in contact with the tissue, the stress due to the thermoelastic wave modulates its thickness and, hence, the

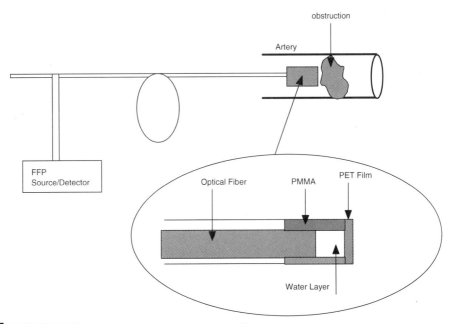

FIGURE 8.25 Angiology instrument using optical fibers.

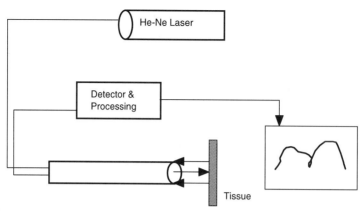

FIGURE 8.26 The fiber-optic Doppler flow meter.

optical phase difference between the interfering Fresnel reflections from both sides of the film.

When the sensor head is coaxially positioned with the delivery fiber, measurements are taken from the center of the acoustic source, giving improved targeting accuracy. This kind of photoacoustic spectroscopy has been experimentally tested on postmortem human aortas.

In addition to disturbing central circulation, cardiovascular diseases may influence the peripheral circulation by affecting the microvascular perfusion in tissue. Insufficient peripheral circulation may produce chemical gastric discomfort and ulcers. The best method for assessment of microvascular perfusion is laser Doppler flow monitoring in which the use of optical fibers can improve the possibilities of both invasive and contact measurements.

The basic concept of fiber-optic laser Doppler flow is shown in Figure 8.26. Light from the He-Ne laser is guided by an optical fiber to the tissue or vascular network being studied. The light is diffusely scattered and partially absorbed within the illuminated volume. Light hitting a moving blood cell undergoes a small Doppler shift due to the scattering particles.

Gastroenterology

The need for fiber optic systems to monitor *in vivo* the functional aspect of the foregut is increasing. An important parameter when studying the human foregut is the gastric and esophageal pH. Monitoring gastric pH for long periods serves to analyze the physiological pattern of acidity. It provides information regarding changes in the course of the peptic ulcer and enables assessment of the effect of gastric antisecretory drugs. In the esophagus, the gastroesophageal reflux, which causes a pH decrease in the contents from 7 to 2, can determine esophagitis with possible strictures and Barrett's esophagus, which is considered a pheneoplastic lesion. In addition, in measuring the bile-containing reflux, the bile and pH should be measured simultaneously.

The sensor shown in Figure 8.27 uses two dyes. Chromophores, immobilized on controlled pore glasses, are fixed at the end of plastic optical fibers. The distal end of the fibers is heated and the CPG forms a very thin pH-sensitive layer

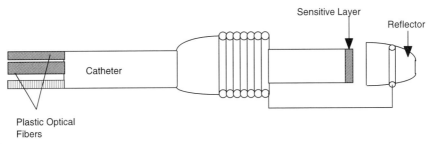

FIGURE 8.27 Gastroenterology usage of fiber optics.

on the fiber tips. The probe has four fibers (two for each chromophore). The use of LEDs as light sources, solid-state detection, and an internal microprocessor makes this a truly portable, battery-powered sensor.

REFERENCES

Bruce, R. V. 1973. *Alexander Graham Bell and the Conquest of Solitude*. Gollanez, London.
Cherin, A. H. 1983. *An Introduction to Optical Fibers*. McGraw-Hill, New York.
Hall, R. N., G. E. Fenner, J. D. Kingsley, T. J. Soltys, and R. O. Carlson. 1962. "Coherent light emissions from GaAs Junction." *Phys. Rev. Lett.* 9, 366.
Kao, K. C., and G. A. Hockham. 1966. "Dielectric fiber surface waveguides for optical waveguides." *Proc. Inst. Electr. Eng.* 133, 1151.
Kapron, F. P., D. B. Keck, and R. D. Maurer. 1970. "Radiation losses in glass optical waveguides." *Appl. Phys. Lett.* 17, 423.
Maiman, T. H. 1960. "Stimulated optical radiation in ruby." *Nature* (London) 6, 106.
Morse, P. M., and H. Feshbach. 1953. *Methods of Theoretical Physics, Part II*. McGraw-Hill, New York.
Nathan, M. I., W. P. Dumke, G. Burns, F. H. Dill, Jr., and G. Lasher. 1962. "Stimulated Emissions of Radiation from GaAs p-n Junctions." *Appl. Phys. Lett.* 1, 62.

INDEX

Accelerometers, 102–104
Aliasing, 194–196
Alvarez structure, 63
Amplifiers
　See also Noise, in operational amplifiers
　chopper-stabilized, 88
　current feedback, 148–149
　instrumentation, 86–88
　isolation, 89
　signal-conditioning, 78, 84–86
Analog signals, discrete time sampling of, 194–196
Analog-to-digital converter (ADC)
　coherent and noncoherent sampling, 188–191
　digitally corrected subranging, 135, 141
　driving, with switched capacitor inputs, 118–120
　example of, 109–110
　external protection of amplifiers, 131–134
　external reference voltage generation, 122–123
　flash, 135–138
　gain setting and level shifting, 120–122
　high-speed architectures, 135–141
　high-speed sampling, 112–116, 122–123
　input protection, 123–124
　interfacing, to digital signal processors, 214–217
　multichannel acquisition, 111–112
　multichannel applications, 128–131
　noise issues, 124–128, 157–159
　parallel interfacing ADC-to-DSP, 217–219
　sample and hold conversion, 110–111
　selection of drive amplifier for, 116–118
　serial interfacing ADC-to-DSP, 225–227
　subranging, 134, 139–140
　successive approximations, 135, 138–139
Anderson, Weston, 34–35
Antiparallel state, 46
Arithmetic logic unit (ALU), 205, 207–208
Arithmetic status register (ASTAT), 212
Arrays, charge-coupled device, 240–242
ASIC
　ground bounce, 176–179
　parasitic extraction, 181–183

Ball-grid array (BGA), 29
Betatrons, 57–58
Bias current, 151
Bin size, 186
Biquad, 201, 202
Blackman–Harris function, 188–189
Bloch, Felix, 33, 45
Block equations, 36, 49
Boltzmann's constant, 42, 48
Boltzmann's law, 48
Bridges, 79–80, 81

275

driving, 80, 82–83
Brokaw cell, 106
Brownout, partial, 21
Burst noise, 148
Bypass capacitors, 159–161
   resonances and, 162–166
   using two or more, 166–168

CAD programs, 200
Capacitive charge transducers, 95
Charge-coupled devices (CCDs), 95–98
   arrays, 240
   description of, 237–242
   interline transfer, 240–242
Charge-emitting sensors, 95
Chemical shift, 39
Chopper-stabilized amplifiers, 88
CODECs to DSP, interfacing, 229–233
Coherent sampling, 188–191
Color noise, 143
Common mode choke, 11, 134
Common mode interference, 9–11, 20–21
Common mode rejection ratio (CMRR), 86, 87, 146
Complementary metal-oxide semiconductor (CMOS), 29
Compton scattering, 260
Computer tomography (CT) scanners
   development of, 259–261
   digital imaging, 263
   sectional imaging, 261–263
Conducted interference, 9, 10
Contact image sensor (CIS), 95
Coolidge, W. D., 57
Cormack, Allan, 261
Corner frequency, 147
Crosstalk through PC card pins, 179–181
Current feedback amplifier, 148–149
Cutoff condition, 70

Data acquisition. *See* Analog-to-digital converter (ADC)

Data address generators (DAGs), 211–212
Defibrillators
   high voltage, 28–30
   implantable cardiovascular (ICDs), 29
Dickinson, W. C., 34
Digital filters
   digital signal processing techniques for, 196–197
   finite impulse response, 197–200
   infinite impulse response, 200–202
Digitally corrected subranging ADCs, 135, 141
Digital signal processing/processors (DSP)
   applications, 234–236
   arithmetic logic unit, 205, 207–208
   hardware, 205–207
   interfacing ADCs and DACs to, 214–217
   interfacing I-O ports and CODECs to DSP, 229–233
   multiplier-accumulator, 205, 208–210
   parallel interfacing ADC to, 217–219
   parallel interfacing DAC to, 220–223
   serial interfacing, 223–225
   serial interfacing ADC to, 225–227
   serial interfacing DAC to, 227–229
   shifter, 205, 210–211
   summary of interfacing, 233–234
   techniques, 196–197
Digital-to-analog converter (DAC)
   interfacing, to digital signal processors, 214–217
   parallel DAC-to-DSP, 220–223
   serial DAC-to-DSP, 227–229
Digital x-rays, 264–268
Discrete Fourier transform (DFT)
   See also Fast Fourier transform (FFT)

   calculation of, 185–186, 202–204
Discrete time sampling of analog signals, 194–196

Electromagnetic interference (EMI), 3, 8–12
Electron beams, applications, 59–60
Electrostatic discharge (ESD), 133
Endoscopes, 263–264
Ernst, Richard, 34–35

Faraday cages, 11, 12
Faraday screens, 19
Fast Fourier transform (FFT)
   coherent and noncoherent sampling, 188–191
   determining time-domain samples and record length and performing, 186–188
   role of, 185, 202–204
   ultrasound application, 191–194
Fast Fourier transform (FFT), hardware implementation, 204
   arithmetic logic unit, 205, 207–208
   data address generators (DAGs), 211–212
   digital signal processing hardware, 205–207
   multiplier-accumulator, 205, 208–210
   program sequencer, 212–213
   sampling rates, 205
   serial ports, 213–214
   shifter, 205, 210–211
   system interface, 214–234
Feedback, current versus voltage, 150
Feedback factor, 147
FET converter, two-transistor, 24–26
Fiber optics
   advantages of, 245–246
   classification and features of, 244–250
   communications, development of, 242–244

graded-index, 245, 257–259
loss mechanisms, 247
medical sensors from, 268–273
step-index, 245, 251–257
Filter inductors and capacitors, 6, 8
Filters, output, 19–21
Finite impulse response (FIR) digital filters, 197–200
Flash converters, 135–138
Flicker noise, 148
Flightback switch mode power supplies, 22–24
Forward converters, 26–28
Fourier transform technique, two-dimensional, 55
Full-power bandwidth (FPBW), of op-amps, 153

Gain-bandwidth (GBW) products, 153–154
Gain setting, 120–122
Graded-index fiber optics, 245, 257–259
Ground bounce, 176–179

Half-bridge flyback converters, 24–26
Hall effect magnetic sensors, 100–101
Hamming function, 185, 188
Harmonic distortion, 186
Hounsfield, Godfrey, 261

IEC regulations, 8–9
IEEE standard 587-1980, 3–5
Infinite impulse response (IIR) digital filters, 200–202
Input offset voltage, 151–152
Input protection, ADC, 123–124
Inrush current control 12–13
Instrumentation amplifiers, 86–88
Insulated-gate bipolar transistors (IGBTs), 29
Interfacing
 ADCs and DACs to digital signal processors, 214–217
 I-O ports and CODECs to DSP, 229–233

parallel ADC-to-DSP, 217–219
parallel DAC-to-DSP, 220–223
parallel to DSP processors, 22
serial ADC-to-DSP, 225–227
serial DAC-to-DSP, 227–229
serial to DSP, 223–225
summary of, 233–234
Interline transfer, 240–244
Interrupt control register (ICNTL), 212, 213
Interrupt force and clear register (IFC), 212
Interrupt mask register (IMASK), 212
Inverting case, 150
I-O ports to DSP, interfacing, 229–233
Isolation amplifiers, 89

Johnson noise, 84, 85–86, 124, 143

Larmor resonant frequency, 42, 48, 52
Lauterbur, Paul, 35
Level shifting, 120–122
Linear RF accelerators, 59, 63–64
Linear variable differential transformers (LDVTs), 98–100
Loads, protection for linear and nonlinear, 16–17

Magnetic resonance imaging (MRI)
 applications, 40
 development of, 33–37
 hardware design, 50–53
 high-resolution, 39
 how it works, 41–50
 pulses, 52
 review of, 37–40
 spin echo pulsing, 53–55
 two-dimensional, 35, 55
 zeugmatography principle, 35
Magnetrons, 69–70
Magnets, MRI permanent, 52
Medical Devices Directive, 8

Medical sensors, from fiber optics, 268–273
Metal oxide semiconductors (MOSs), 237
Metal-oxide varistors (MOVs), 6
Metal semiconductor capacitors, 237
Miller current, 19
Mode status register (MSTAT), 212
Motion sensors, 98–104
Multichannel acquisition, 111–112
Multiplier-accumulator (MAC), 205, 208–210

Noise
 crosstalk through PC card pins, 179–181
 designing power bus rails to control, 168–171
 ground bounce, 176–179
 in high-speed sampling ADCs, 124–128, 157–159
 parasitic extraction, 181–183
Noise,
 in operational amplifiers (op-amp), 84–86
 bypass capacitors, 159–168
 bypass capacitors and resonances, 162–166
 calculation of, 143–146
 fundamental specification, 146–151
 gain-bandwidth products, 153–154
 input offset voltage, 151–152
 internal, 154–156
 noise gain, 152
 power supply decoupling methods, 159–162
 slew rate and full-power bandwidth, 152–153
Noncoherent sampling, 188–191
Noninverting case, 150
Nuclear magnetic resonance (NMR). See Magnetic resonance imaging (MRI)
Nutation, 37
Nyquist rate, 110, 194

Optical encoders, 102
Optical fibers. *See* Fiber optics
Optical sensors
  charge-coupled devices, 237–242
  CT scanners, 259–263
  digital x-rays, 264–268
  endoscopes, 263–264
  fiber optics devices, 242–257, 268–273
  graded-index fiber, 257–259
Optoisolators, 91
Output filtering, 19–21
Overload protection, 16–17
Overvoltage protection, 14, 15

Parallel interfacing
  ADC-to-DSP, 217–219
  DAC-to-DSP, 220–223
Parasitic effects, 20
Parasitic extraction, 181–183
Particle accelerators
  advantages and disadvantages of, 59
  applications, 62
  architecture, 70–73
  development of, 57–59
  electron beams, 59–60
  hardware, 67–68
  linear RF, 59, 63–64
  magnetic fields and, 64–65
  magnetrons, 69–70
  properties, 60–61
  standing wave system, 71
  synchrotron, 66–67
  traveling wave system, 71
Particle magnetic rigidity, 64
PC card pins, crosstalk through, 180–181
Photodiodes, 78, 93–94
Piezoelectric force transducers, 90–91
Plank's constant, 42, 47
Popcorn noise, 148
Position sensors, 98–104
Power bus rails, to control noise, 168–171
Power failure warnings, 21–22
Power spectral density, 143
Power supply, designing
  electromagnetic interference, 3, 8–12
  flightback switch mode power supplies, 22–24
  forward converters, 26–28
  half-bridge flyback converters, 24–26
  high voltage defibrillators, 28–30
  inrush current control 12–13
  output filtering, 19–21
  overload protection, 16–17
  overvoltage protection, 14, 15
  power failure warnings, 21–22
  snubber circuits, 17–19
  soft start circuit, 13–14
  transient voltage protection, 3–8
  undervoltage protection, 14–16
Power supply decoupling methods, 159–162
Power supply rejection ratio (PSRR), 160–161
Proctor, Warren, 34
Program sequencer, 212–213
Pulse Fourier transform, 34–35
Pulses, MRI, 52
Purcell, Edward M., 33, 45

Rabi, I., 33
Radiated interference, 9, 10
Rectifier capacitor input filter, 12, 13
Resistance temperature detectors (RTDs), 78
Resistive elements, 79
Resonance frequency, 42
Resonances, bypass capacitors and, 162–166
RF fields, 52
RF linear accelerators, 59, 63–64
Rontgen, Wilhelm, 259

Sample and hold conversion, 110–111
Sensing, physical principles of, 78–79
Sensors
  active, 78
  characteristics of, 75–78
  charge-coupled device, 95–98
  chopper-stabilized amplifiers, 88
  driving bridges, 80–83
  high-impedance, 93–94
  instrumentation amplifiers, 86–88
  interfacing, 79–80
  isolation amplifiers, 89
  passive, 78
  position and motion, 98–104
  signal-conditioning amplifiers, 78, 84–86
  strain, force, pressure, and flow, 90–93
  temperature, 104–107
Serial interfacing
  ADC-to-DSP, 225–227
  DAC-to-DSP, 227–229
  to DSP, 223–225
Serial ports, 213–214
Shifter, 205, 210–211
Shim coils, 51
Shot noise, 148
Signal-conditioning amplifiers, 78, 84–86
Slew rate, of op-amps, 152
Snubber circuits, 17–19
Soft start circuit, 13–14
Spin, physics of
  packets, 43
  properties, 41
  T_1 processes, 43–44
  T_2 processes, 44–50
  transitions, 41–43
Spin echo pulsing, 53–55
Spin-spin relaxation time, 44
Spin-warp imaging, 36
Stack status register (SSTAT), 212
Step-index optical fibers, 245, 251–257
Strain gauges, 78, 90
Subranging ADCs, 135, 139–140
Successive approximation ADCs, 135, 138–139
Switched capacitor inputs, 118–120
Synchrotron, 66–67

Temperature sensors, 104–107
Thermal noise, 144–146
Thermistors, 78, 104
Thermocouples, 78, 104–105

Trace resistance, 171–176
Transient protection diodes, 6
Transient suppression diodes, 6, 7

Ultrasound application, 191–194
Undervoltage protection, 14–16

Van de Graff, R., 57
Varistors, 6
Velocity modulation, 67

Voltage
  feedback, 150
  input offset, 151–152
Voltage protection, 3–8
  overvoltage, 14, 15
  under, 14–16

Weighting function, 185, 188
Wheatstone bridge, 79, 80
White noise, 143
Windowing, 187, 188, 191

WKBJ method (Wentzel, Kramers, Brillouin, and Jefferies), 257–258

X ray technology, 259–261
  See also Computer tomography scanners
  digital, 264–268

Yu, F. C., 34

Zeeman effect, 45

R
856
.P465

2002